GROVER E. MURRAY
STUDIES IN THE
AMERICAN SOUTHWEST

A Haven in the Sun

Five Stories of Bird Life and Its Future on the Texas Coast

B.C. Robison

Illustrations by Linda M. Feltner

TEXAS TECH UNIVERSITY PRESS

This book is typeset in Cormorant Garamond. The paper used in this book meets the minimum requirements of ANSI/NISO Z39.48-1992 (R1997). ∞

Designed by Hannah Gaskamp
Jacket illustration by Linda M. Feltner

Generously supported by the Houston Museum of Natural Science

Library of Congress Cataloging-in-Publication Data

Names: Robison, B. C., 1946– author. | Feltner, Linda M., illustrator.
Title: A haven in the sun: Five stories of bird life and its future on the Texas coast / B. C. Robison; with illustrations by Linda M. Feltner.
Description: Lubbock, Texas: Texas Tech University Press, [2020] | Includes bibliographical references and index. | Summary: "A natural history work, with illustrations, written for the general reader, providing an account of birds--Attwater's Prairie Chicken, White-tailed Hawk, Whooping Crane, Redhead, and migratory shorebirds and songbirds--of the Texas Coast"-- Provided by publisher.
Identifiers: LCCN 2020003802 | ISBN 9781682830635 (cloth)
Subjects: LCSH: Birds--Texas--Gulf Coast.
Classification: LCC QL684.T4 R63 2020 | DDC 598.0976--dc23
LC record available at https://lccn.loc.gov/2020003802

Printed in the United States of America
20 21 22 23 24 25 26 27 28 / 9 8 7 6 5 4 3 2 1

Texas Tech University Press
Box 41037
Lubbock, Texas 79409-1037 USA
800.832.4042
ttup@ttu.edu
www.ttupress.org

To Middy Randerson,
whose love and support have been the joy of my life,
and
to the memory of Walter Harrison Skeen,
who showed me my first American Oystercatcher,
on Fulton Beach Road

Even the sparrow has found a home,
And the swallow a nest for herself,
Where she may lay her young . . .

Psalm 84:3

Contents

List of Illustrations

ILLUSTRATIONS BY LINDA M. FELTNER

TEXAS

Attwater Prairie
Chicken N.W.R.

Eagle Lake

Victoria

Goliad

Tivoli

Aransas
N.W.R.

Matagorda
Peninsula

Matagorda
Island

Fulton
Rockport

Blackjack Peninsula

Aransas Pass

San José Island

Corpus Christi

King Ranch

Upper Laguna Madre

Baffin Bay

Padre Island

Kenedy Ranch

Land Cut

0   10   20 miles
North 10   20   30 kilometers

Lower Laguna Madre

Mexico

Rio Grande

Boca Chica

Brownsville

Beaumont

Louisana

Sabine Pass

High
Island

Galveston
Bay

Bolivar
Peninsula

Bolivar Flats
North Jetty
South Jetty

Gulf of Mexico

LINDA M FELTNER

A Haven in the Sun

Introduction

The Texas Coast
Land, Water, Birds, People

IT TOOK MANY YEARS for me to become captivated by the Texas Coast, and even longer to appreciate its bird life. Growing up in southeast Houston, I looked upon the beach as a place I had to be while my neighborhood buddies back in Golfcrest — Glenn, Lee, Jimmy, and Jack — were riding bicycles along Telephone Road. Each summer my mother would drag me down to visit Aunt Quillian at her wretched beach house in Surfside, then a scruffy community south of Galveston. Aunt Quillian was a tall, cranky old spinster who painted what I think were landscapes, dabbled in genealogy, wore earrings that could be used as fishing lures, and had a home in Angleton whose interior was inspired by a South American bat cave. She was the first non-parent to yell at me, when I asked, in a moment of seven-year-old audacity, how old she was. "Why, that's none of your business!" she shrieked, glaring down at me through her tortoise-shell glasses. "That old, huh?" I replied. Well, I didn't really say that, but I wanted to.

I longed to be elsewhere. The grown-ups talked for hours about long-forgotten cousins while I sat around, bored and sulking. The water, pumped from a rusty, hand-cranked pump, was a ghastly, briny bilge. The house was hot and the beds were hard. I would sneak outside to escape the tedium and found nothing but hot sand, baking heat, prickly plants, and biting bugs. The inviting surf, a few dozen yards away, was forbidden territory.

Going to the coast meant visiting Aunt Quillian, and I didn't like either it or her.

Then there were the endless weekends on Bolivar Peninsula, where my father, a real estate developer, was building the communities of Emerald Beach and Ramada Beach. I stood around watching the workers as they sawed lumber, pounded pilings, painted fences, bulldozed roads. He had this crew of shifty characters from Louisiana working for him — guys whose recent addresses, shall we say, reflected a certain unavailability for the past few years. One guy was missing a finger and would never tell me what happened to it. I stood around, bored and fidgety, listening to the hokey music on the radio — *Vooolaaaare, oooooh oh!* — and fretting that Glenn and the guys were at the bowling alley or the neighborhood swimming pool without me.

One thing — make that the only thing — that I enjoyed about going to Bolivar in those days was riding the ferry from Galveston. There was somehow a comforting routine to it all, and it was exciting — the *clang! clang!* of the landing ramp as cars drove on board, the crew sternly directing you to your spot, the sway of the vessel as cars went to one side, then to the other, the low growl of the idling engines, the jarring blast of the horn and the growing rumble as we pulled away, the three-mile trip across Galveston Bay dodging tankers and fishing boats, the crying gulls bouncing over the wake, the *whomp!* upon settling into the landing on the opposite side, followed by a final *clang!* on the ramp as we drove off. It is an experience that I enjoy to this day.

My father occasionally took me to a beer joint (he knew every hangout on Bolivar), where he would have a Falstaff and I would have a hamburger. The local characters — shrimpers, fishing guides, unshaven residents — were a study in advanced raffishness. One time, this old gent, who apparently was into his seventh or eighth beer, was regaling everyone in earshot about the fact that he had been married and divorced thirteen times. This went on for many minutes, to everyone's annoyance. Finally, my father, never one to suffer fools gladly, yelled to the guy: "Are you bragging or complaining?" Sunday afternoons would mercifully come to an end and we would go home.

I began to enjoy the coast when I started spending summers at my grandmother's house, in Corpus Christi, where my mother was born a few months before the 1919 hurricane that struck the city. Mother would take me to Union Station in downtown Houston, tuck a five-dollar bill in my pocket, and put me on Missouri Pacific's Valley Eagle for the trip to, as the gentlemanly old porters announced, "Corpus Chrischa." The trip took forever, but I loved it: strolling through the echoing terrazzo cavern of the elegant station, getting the rare luxury of a shoe shine, the feeling of grown-up-ness from walking down the platform with slicked-back hair and a small suitcase, feeling the anticipation of boarding the humming railcar with its chilled metal-scented air, sitting high up in the seat as we trundled out through the small farming communities southwest of Houston — Wharton, Ganado, Edna — eating a fancy sandwich in the dining car, and, finally, hours later, chugging into Corpus Christi's Broadway Street station.

Granny — my mother's mother — was a beloved lady who lived in an airy, comfortable two-story frame house on King Street, near South Bluff Park, not far from downtown and the waterfront. She lived there with her second husband; my mother's father died when she was fourteen. Aunt Bonnie and Uncle Walter, my mother's brother, lived in a small garage apartment in the back with their daughters, my two young cousins Saralee and Molly.

Granny's husband sat all day in a white wicker rocking chair with the *Caller-Times* on his lap. I never saw him do anything around the house and I always had the feeling he didn't quite know who I was. Family lore recounted that, as a young man working on the Southern Pacific Railroad, he pulled into Galveston one Friday afternoon in early September 1900. The weather was unsettled, but apparently the citizens of Galveston were unconcerned. As the weather became violent overnight, he supposedly tied himself to a tree, surviving the now-famous hurricane. Another part of the story was that the only thing left on him afterwards was his work boots, an image which I never cared to dwell on.

It was in Corpus Christi that I began to have some vague notion of this thing called "nature." Granny's house was not air-conditioned, so the windows were usually open, letting in the warm, fragrant breeze and the sound of buzzing insects during the day. Behind the house and right outside my bedroom window stood a small orchard of fig trees. Late one night I heard the snuffling and rustling of some creature rooting in the large, raspy leaves on the ground. The noise went on and on, and after a while it became a little scary. I finally got the nerve to look out the window and there it was, this strange animal with sharp teeth and glowing eyes in the silvery moonlight: possum! It poked around the leaves and then wandered off. He — she? — returned periodically for a night visit, which I observed thereafter with less apprehension.

Figs fell off the trees, attracting butterflies and, I guessed, my night-time visitor. Dozens of butterflies perched around the rotting fruit during the day; I had never seen so many of these dancing, shining creatures so closely. My wonderful Aunt Bonnie, who had studied at the University of Iowa, taught me my first big scientific word — "lepidopteran" — the taxonomic term for butterflies and moths. Uncle Walter would take me down to Aransas Pass and Rockport-Fulton. "Seagulls" were everywhere. Small gray birds skittered along Fulton Beach Road but I didn't pay them much attention. They all looked the same.

Everything seemed different in Corpus Christi — the hazy sunlight, the everlasting wind, the tall palms along Ocean Drive swaying in the fish-tangy bay-breeze, the laughter of the gulls on the waterfront, the casual small-town feel. I walked for hours around downtown or along South Staples Street, something a ten-year-old could safely do in that sleepy Eisenhower era. I went to movies at the Amusu, browsed through books at La Retama Library, sat around the T-head marinas on the bay, talking to Mexican fishermen who taught me Spanish words. I prowled around the gently undulating landscape of South Bluff Park, with its dark, gnarled mesquite groves and little Art Deco museum built for the Texas Centennial, in 1936. I visited the beautiful Corpus Christi Cathedral down on the bluff, overlooking the bay, its cool dim candle-scented silence a welcome relief from the summer heat. Other adventures awaited with Aunt Bonnie and Uncle Walter: trips to Padre Island, dinners at dimly-lit Tex-Mex joints out on Agnes Street, visits to the Naval Air Station, which in those days you could just drive into with a wave from the guard; now you can't even get within half a mile of the entrance.

Back home in Golfcrest, I had a semi-arboreal childhood with a huge black willow tree in our spacious back yard. My friends and I climbed it almost every day in the summer and we had a tire on a rope. When the tree died of dry rot years later, it was like a death in the family. My father put up a bird bath and feeders out back and let vines and shrubs grow wild and untended (a family tradition I maintain today). Jays, grackles, sparrows, crows, and cardinals flitted and fed around this little piece of urban wilderness. I asked my father one day why the back yard was such a mess; he said that birds need "habitat," patiently concealing his exasperation with my apparent lack of any sense of ecology. And we always had dogs and cats at home. We never had many at one time, but a stray would appear in the yard, decided she liked the accommodations, and proceeded to live a life of care and comfort with us for a dozen years or so. Then another would arrive and adopt us.

I first became aware of birds being an object of special interest from my Aunt Norma. On the way to visit "Aunt Nor'" one afternoon at her home in Houston's West University, my mother said, in what I think may have been hushed tones, "You know, Aunt Nor' watches birds as a hobby." The significance of this news did not register with me immediately, but I did have the uneasy feeling that I should not tell anyone at school about it. I queried Aunt Norma about this strange pastime.

"Well, I go out in the woods and look at birds," she said.

"And then what?" I asked.

"I identify them," she answered. "Add them to my list."

When pressed as to what this "list" was for, she explained that it was a "life list," a list of all the birds you have ever seen. Well, OK, I guess. Norma C. Oates had a long life list.

She could have been a poster girl for the birdwatcher of popular imagination: literally, a little old lady in tennis shoes, with a helmet of short gray hair, rimless glasses, and battered binoculars around her neck. I discovered years later when I took up birding myself that my kindly, quick-witted aunt was well known in birdwatching circles. In fact, she has a citation in the definitive work on Texas ornithology, H. C. Oberholser's *The Bird Life of Texas*, for being the first to report nesting Cattle Egrets in Texas, in Galveston in 1958. She was friends with Connie Hagar, the famed birder and conservationist from Rockport.

She had a shelf of bird books in a back room at her house where I escaped to avoid boring relative-talk. The books had weird titles — *Trogons of Hispaniola* or whatever — but they told me that Aunt Norma was quite serious about this bird business. I remember a big picture book with birds in all these strange, unsightly contortions, especially large birds like cranes and herons. Their necks were twisted down in painful-looking ways and they appeared all out of proportion. "This guy can't draw," I recall thinking. The book was *The Birds of America* by John James Audubon.

My youthful inability to appreciate America's greatest wild-life artist raised its ugly head years later when Houston's Museum of Fine Arts held an exhibit of original lithographs of Audubon's birds of prey. The hawks, owls, falcons, kites, and eagles — painted life sized from propped-up specimens, as Audubon famously created his birds — glared at me with a stunning and disquieting menace; they appeared to be on the verge of flying out of the frame and pouncing on my head. The vitality, the harnessed energy, the precision of Audubon's birds render the works of the country's other important early wildlife artists, Mark Catesby and Alexander Wilson, flat and lifeless.

Besides dogs, cats, the big willow tree, and my neighborhood friends, the other great joy of my childhood was books. My parents never had much extra money, but they did allow themselves the luxury of a membership in the Book of the Month Club and a subscription to *National Geographic*. We had musty old mid-century titles all over the house, from authors like Bernard DeVoto and James Thurber, along with stacks of the familiar yellow-bordered magazine. Seeing my parents reading most nights probably influenced me as much as the books themselves.

My interest in books and nature eventually converged soon after my graduation from the veterinary college at Texas A&M University. I began a campaign of reading classics in natural history literature: Rachel Carson's *The Edge of the Sea* and *The Sea Around Us*; *Life and Death of the Salt Marsh* by John and Mildred Teal; *Sand County Almanac* by Aldo Leopold; Henry Beston's *The Outermost House*; *The Immense Journey* by Loren Eiseley; and the book that had a greater influence on me than any other, Lewis Thomas's *The Lives of a Cell*, a contemplation about man and nature. I even made it through Darwin's *The Origin of Species*, having first come across it years earlier at La Retama Library and not understanding a word of it.

I eventually decided I needed a nature-related hobby, something to get me outdoors and away from the veterinary clinic. Birdwatching seemed to be the logical choice. Birds were fun and interesting to watch, you didn't need a lot of fancy equipment, and you could do it in your back yard. The fact that Houston was situated within one of the most bird-rich regions on the continent was a revelation that came later. Not sure of the best place to begin, I sought advice from Aunt Norma.

"How about Galveston?" I asked.

"Galveston would be a good place to start," she said.

So off I went. I bought binoculars, a spotting scope, and a copy of Roger Tory Peterson's *A Field Guide to the Birds of Texas and Adjacent States*. It was at the foot of the causeway leading to Galveston, in a beautiful marsh of *Spartina* grass, where I saw my first "official" bird: a Louisiana Heron, skulking in shallow water along the grassy edge. A few years later the American Ornithologists' Union, the group that determines the taxonomy of birds, changed the bird's name to "Tricolored Heron," which I suppose is more descriptive but less evocative. Every few years the AOU (now known as the American Ornithological Society) changes the name of some birds, for no immediately apparent reason. I have always suspected they take this step just to have something to do when things get slow around the office, but in fairness let's assume that a more substantial rationale is involved.

Other trips up and down the Texas Coast soon followed: Bolivar Flats and High Island, Anahuac, Eagle Lake for the prairie chickens, the Aransas National Wildlife Refuge, South Texas, Padre Island, and the wonderful wildlife refuges of the Rio Grande Valley, Laguna Atascosa and Santa Ana. I joined the Ornithology Group of the Houston Outdoor Nature Club and went on Christmas Bird Counts.

Those early days of birdwatching are magical, when everything is new and life species come to you with ease. The little birds running around the beach weren't all the same, they were a dozen different species of sandpipers and plovers. You could see seven species of ducks in one pond. Two dozen species of sparrows flitted throughout coastal grasslands in winter. Six commonly occurring species of gulls and eight species of terns glided over the surf. There were long-legged waders and curlews in the marsh and godwits and dowitchers on the mudflats. Many species came and went with the seasons, while others remained year-round. Every birding trip to the coast became a venture of discovery.

I had the privilege of getting to know some of the most accomplished Texas birdwatchers of that era: Ben Feltner, Ted Eubanks, Gary Clark, Noel Pettingell, Bob Behrstock, Jim Morgan. I never came within half a light-year of being as good as they were, but I learned a lot. (The best lesson they gave me: Study the field guide *before* you go out; take it with you but leave it in the car.) I went birding with friends to other good bird areas in the state — the Big Thicket, the Hill Country, Big Bend — but it was the coast, with its marshes and surf and sea breeze, that for me always held the greatest allure.

Around the time I began birdwatching I noticed a weekly column about nature that had been running in the *Houston Chronicle*. It was titled "Nature Trails" and was written by John Tveten, a freelance nature writer and photographer who lived in Baytown, near Houston. Each week he wrote about birds, or turtles, or a state park, among many other topics. John was a complete naturalist: He could tell you the name of a tree in a pasture, the name of the bird in the tree, the name of the wildflower under the tree, and the name of the insect feeding on the wildflower. John passed away in 2009, a dear friend and mentor to many of us.

The idea eventually occurred to me that I would like to do something similar for the city's other daily newspaper, *The Houston Post*. I had been a writer and editor for various publications all through high school, college, and veterinary school and decided I would like to get back into writing.

I called John and asked him how he got started. He graciously explained that he sent some sample columns to the editor with a proposal to do a weekly column. And he then added: "But it took about a year for the editor to decide to run it." This did not bode well for my natural tendency toward impatience, but I thought I would try it anyway.

I sent three columns to Kuyk Logan, the *Post*'s managing editor. About a week later he wrote back, saying, in so many words, "Let's do it." Thus was born the "Texas Naturalist" column, and it ran weekly for almost ten years. My first column featured the green anole, that little green lizard (an iguana, not a chameleon) that you see around the house in summer, changing colors from brown to green. I had fun with the "Texas Naturalist," writing on almost anything about nature that I wanted, including controversial issues like wetlands preservation and the protection of endangered species. I would write an occasional column on evolution, mainly to rile up the creationists.

At that time, the *Post* was a feisty newspaper that took on topics that the conservative *Chronicle* avoided for fear of offending some Chamber of Commerce type. I would see John at bird meetings and he would ask in amazement and, I do believe, with a touch of envy: "How do they let you get away with that?" after I had blasted some environmentally destructive scheme, usually something concocted by the Corps of Engineers. "The *Chronicle* would never let me do that!" John would say. (It should be noted that the *Chronicle* today is a much more progressive newspaper.)

I did occasional pieces for the *Post*'s op-ed page, including a multi-part series on the Wallisville Dam controversy, for which I won the Houston Sierra Club Environmental Reporting Award. This was an extremely contentious project of the Corps much desired by rice farmers in Liberty County to keep salt water out of their rice fields. The scheme, however, threatened Galveston Bay — the state's most economically valuable estuary — with severe ecological damage by reducing freshwater inflow from the Trinity River. Facing vigorous opposition from private citizens, environmental groups, shrimpers, and oystermen, the dam was never built.

Not long after starting the "Texas Naturalist," I retired from veterinary practice and started graduate school in the biology department at Rice University to pursue a career in environmental consulting. Rice had at that time a small university press that published works mostly by faculty members. One afternoon I called Sue Fernandez, the editor, and proposed a field guide for the common backyard birds found throughout Houston. She said, in so many words, "Let's do it." I was on a roll; this publishing business was fun. So for the next year I spent evenings at home writing descriptions of fifty-four common city birds — cardinals, mockingbirds, blue jays, woodpeckers, owls, and others — as my white bob-tailed cat Charlie Rabbit kept me company on my desk.

Birds of Houston, with photographs by John Tveten, was published with generous assistance from the Houston Museum of Natural Science; it served as the first in a series of natural history books published by Rice. The next book in the series was John's excellent guide, co-written with his wife Gloria, to the butterflies of Houston and Southeast Texas. When the Rice press ceased operations, in 1996, the University of Texas Press took over only two titles from Rice's catalogue for continued publication: the bird book and the butterfly book.

The "Texas Naturalist" years coincided largely with my time at Rice. I had the library within easy reach, and I could take time off for birding trips down to the coast. Alas, however, all good things must pass. When I graduated and went to work full time for an environmental consulting firm, I brought the "Texas Naturalist" to an end. I greatly enjoyed writing the column, going on trips to the tropics, roaming the coast looking at birds, and receiving mail from readers. Most of the mail was complimentary, except for the time when I proposed (in semi-jest) making the Whooping Crane the state bird. This generated a firestorm of protests. The entire "Letters to the Editor" section the following week was devoted to angry correspondents questioning my credentials and intelligence. I learned a hard lesson: The Northern Mockingbird has a huge following.

For all its satisfaction, nature writing can be a melancholy endeavor. Peering over the writer's shoulder, like a gargoyle on a cathedral, is the constant specter of vanished wildness, of beauty and mystery forever lost. Wildlife continues to retreat on many fronts, trees and open space disappear, the modern world crowds out the natural. What can the nature writer do?

We can, as a first effort, document the diminishment of wild things. But we have a larger purpose. We wish to influence the way people think about nature. For without knowledge there can be no understanding, and without understanding there can be no concern about the fate of birds and bays and marshes and all the rest. Within the great army of those who work tirelessly to preserve the natural world — concerned citizens, conservation groups, wildlife managers, and scientists — nature writers play their modest but important role.

Today, that role has become ever more urgent. As bird populations decline and habitats are destroyed, those of us who care about the coast should first ask: Why is the Texas Coast important to bird life, and what threats to it lie ahead?

From Sabine Lake, near the vast wetlands of Louisiana's Cameron Parish, to Boca Chica, the remote beach close to the hot, briny lagoons of Tamaulipas, Mexico, the Texas Coast wraps its 367-mile-long bent arc around the northwestern edge of the earth's fifth-largest sea, the Gulf of Mexico. It is a landscape where rivers run through prairie and woodlands down to the sea, creating the nurturing world of estuaries within the ancient valleys that long ago were flooded by the rising waters of melted glaciers. Long, slender islands of sand lie end to end along the mainland, embracing narrow bays and their shallow waters, marshes, and tidal flats. This is the haven of the birds of the Texas Coast, where avian wildlife gathers throughout the seasons in size and diversity seldom matched anywhere on the continent.

More than 480 species of birds find shelter and sustenance in coastal Texas, as a summer home, a winter home, a year-round residence, a rest stop, a refuge in a storm, or a place simply to fly over on their way to somewhere else. The coast has birds of uncountable abundance and birds of perilous rarity. In autumn ducks and geese migrate south to coastal bays and grasslands and in spring songbirds migrate north as they pass through coastal woodlands. Shorebirds come to the coast for the winter, escaping the harsh weather of the Great Plains and Canada. Hawks gather in the fall as they funnel down from the eastern forests on their way to the tropics. Sandpipers that breed in the Arctic pass though the coast on their way to South America. Warblers and orioles and dozens of other species of songbirds that spend the winter in Venezuela and Ecuador stop along the coast on their way to northern forests.

Other species abound: herons, egrets, spoonbills, storks, and others that wade throughout shallow waters; birds that skulk within marshes, such as rails, bitterns, and gallinules; birds that feed on fish in open waters such as cormorants, anhingas, skimmers, frigatebirds, and pelicans; sparrows, larkspurs, meadowlarks, and others that inhabit grasslands; birds that need woodlands, such as woodpeckers, bluebirds, titmice, owls, thrushes, and wrens; falcons, harriers, kestrels, and other birds of prey that hunt over open lands and forests; gull and terns, dancing over the open gulf; and birds that grace the summer sky, such as nighthawks, kites, swifts, and swallows.

Of the 1,112 bird species recognized in 2018 within the continental United States, the Hawaiian Islands, and Canada by the American Birding Association, almost 60 percent occur in Texas. Of the 650 total species recorded in Texas as of 2019 (second only to California, which has 673 to its credit), three-fourths need Texas coastal habitats for some portion of their annual cycle. The Texas Coast has recorded species from 26 of the 31 orders of North American and Middle American birds as defined by the American Ornithological Society.

During the 2017–2018 Christmas Bird Count, as noted in the annual national survey of birds observed in designated 15-mile-wide circles within a 24-hour period, ten of the top fifty tallies came from Texas (including three of the top five), all of them on or near the coast; the number-one count was the Matagorda County / Mad Island Marsh Count, with 220 species. By any measure, the Texas Coast is a region of exceptional importance to North American bird life.

This pageant of bird life rests on the foundation of the coast's geographic centrality and its diversity of habitats. Texas's south-central location within the US mainland and at the juncture of the Americas has led to its frequent description as a "crossroads" for bird life; this is not just a picturesque cliché but a physical reality. Eastern and western US species as well as temperate and tropical species across North, Central, and South America converge within the state's coastal habitats.

Coastal Texas lies opposite the Yucatan Peninsula and the gracefully curving Mexican coastline on the Bay of Campeche, the main points of departure for migrant songbirds and shorebirds crossing the gulf from the tropics during spring migration. Its proximity to Mexico places it in the path of land-based spring migrants. The coast lies midway within the southern tier of states; Aransas County, in the Coastal Bend, is approximately the same distance from San Diego as it is from Miami. The Central Flyway, the great avian highway for the fall migration of ducks and geese, terminates at the coast. As Roger Tory Peterson points out in the preface to his well-known Texas field guide, birders in the state previously needed two books: his guides to eastern and western birds of the United States. Texas embraces them both.

The dominant features of the coastal zone — the prairie with its rivers and woodlands and the estuaries, bays, and barrier islands with their 1,400 miles of shoreline — create the tapestry of habitats that supports such a diverse community of bird life. Ten of Texas's thirteen major rivers — the San Jacinto, Colorado, and Nueces among them — flow into six of the coast's seven estuarine systems: the Sabine, Galveston–Trinity, Matagorda, San Antonio, Copano–Aransas, and Corpus Christi–Nueces. The Laguna Madre of Texas, a hypersaline lagoon that is the largest of Texas's estuaries, does not receive river flow today, although the Rio Grande connected to it in historic times. Seven barrier islands — Galveston, Follets, Matagorda, San Jose, Mustang, Padre Island, and Brazos (the tiny island south of Padre Island) — and two peninsulas — Bolivar and Matagorda — lie parallel to the coastline and separate the mainland from the gulf, creating elongated bays such as Aransas Bay, East Bay, and Matagorda Bay. Barrier islands and peninsulas lie along most of the coast; they are absent for only about sixty miles of coastline. The Texas chain of barrier islands is one of the longest in the world, and Padre Island is the world's longest single coastal barrier. Channels known as tidal inlets separate the various barriers, allowing water and marine life to circulate between offshore gulf waters and bays. Marshes and tidal flats — broad wind-sculpted areas of sand or mud — lie along the bay margins. Sandy beaches face the gulf.

Annual rainfall declines steeply from the upper to the lower coast, giving the region its distinctive change in landscape; Beaumont's annual rainfall of sixty inches is more than twice that of Brownsville. The reasons for this are numerous and complex, but geography plays a major role. Cold fronts from the west bring in a greater mass of dry air to Brownsville, which is approximately 270 miles south of Port Arthur. The South Texas city lies close to the Mexican Plateau, from which the hot, dry winds blowing north diminish rainfall.

This climatic gradient alters the face of the land as it wraps around the gulf. The upper coast is a humid, temperate landscape with forests and grasslands, brackish estuaries with abundant river flow, and dense marshes along bay shorelines. Between Freeport and Corpus Christi, marshes thin out as the climate becomes hotter and drier. South of Corpus Christi, mesquite, huisache, Texas ebony, and many other trees and plants well adapted to a more arid environment become dominant throughout the countryside.

Another force, however, exerts itself on the life of our coastal birds, beyond geography and climate. With an estimated 2018 population of 28.7 million, Texas ranks second in population in the US, after California. The eighteen counties that make up the coastal zone account for only six percent of the state's area but they are home to a fifth of Texas's population and account for a third of its economic activity. Houston has more than doubled in population since 1960. We should now ask: How can there be enough room for both birds and people in this crowded, busy part of the world? More important, will birds and people be able to live together twenty-five or fifty years from now?

These questions set the framework for the chapters that follow. Birds that have a unique relationship to the landscape of the Texas Coast are the subject of this book: Attwater's Prairie Chicken, the world's most endangered grouse, which has its home only on the coastal prairie; the White-tailed Hawk, found in the US only on the coastal savannah; the Redhead, a duck whose nearly entire population spends the winter on one of the world's rarest bodies of water, the Laguna Madre; the iconic Whooping Crane, whose mid-coastal sanctuary saved it from the ultimate fate of extinction; and migrating shorebirds and songbirds — renowned in the world of birdwatching — passing through each fall and spring on their journeys over the Americas.

These birds not only represent the array of vital habitats that the coast provides — bays, marshes, prairie, woodlands, sandy shores — they also speak to the larger world of Texas bird life and the many challenges it faces for survival. The fate that lies ahead for the birds of this book also awaits many other species of our coastal bird life. There are birds that live in marshes with the Whooping Crane; birds that depend on the prairie beside the prairie chicken; birds that inhabit bays with the Redhead; and many other species that share the coastal savannah with the White-tailed Hawk. Throughout the book one theme prevails: the ways in which people use the land — grass, trees, water, soil — have affected our coastal bird life. The bonds between birds and landscape reveal themselves in unexpected ways in Texas.

As we shall see, the human community has both harmed and nurtured our native bird life. "With large-scale alterations in landscape," says Dr. Richard Gibbons, Conservation Director of the Houston Audubon Society, "there are winners and losers in the world of birds." Perhaps the most striking example of this is the Gulf Coast rice industry. The loss of native prairie to rice farming in the first few decades of the twentieth century largely destroyed the habitat of Attwater's Prairie Chicken. But waterfowl and shorebirds today benefit from those same rice fields, which have replaced tens of thousands of acres of native wetlands lost to agriculture and urban growth.

No part of the Texas Coast has been completely untouched by human activity. While some areas, such as those adjacent to the Laguna Madre and Matagorda Bay, have been blessedly less disturbed than others, most coastal regions in Texas are under continuing pressure from urban and agricultural development. New housing developments spread around the popular Coastal Bend, close to the Aransas National Wildlife Refuge. Wind-energy turbines blight the landscape on the Kenedy Ranch. Water from our rivers — the lifeblood of estuaries — is increasingly in demand for cities and farms. The Katy Prairie continues to yield to highways and neighborhoods. Texas rice farmers have lost more than 450,000 acres of rice lands over the past thirty years to drought and housing development. Almost 20,000 acres of salt marsh were lost between 1990 and 2010. South Padre Island disappeared decades ago under a forest of condominiums.

I was driving around Baffin Bay one late December afternoon when Sandhill Cranes started flying over to their evening roosting place on the King Ranch. I had seen them earlier that day in a wet pasture near the highway. Wave after wave, hundreds of these tall, silver-gray, crimson-capped birds flew over in great arching flocks, their raucous trumpeting music filling the cold twilight. I felt like I was in a cathedral of cranes. A local landowner told me later that wind-energy turbines were being planned for the area.

The natural forces that have altered the coast for millennia — tides, wind, land subsidence, longshore currents, sediment influx, storms, the advance and retreat of the gulf — are joined today by warming atmospheric temperatures and rising sea levels. Depending on location, sea levels have risen five to seventeen inches along the Texas Coast over the past 100 years. Increased temperatures in Texas coastal waters are causing significant changes in distribution of aquatic plants, such as the black mangrove, as well as in fish populations and on the timing of bird migration. Some estimates predict a global seawater rise of one to four feet by the end of this century, endangering coastal cities, with the western Gulf of Mexico expected to endure even greater increases. The effect of rising water on Texas's coastline is compounded by land subsidence, largely from extraction of groundwater and oil and gas. Erosion presents an additional challenge. For years, Texas barrier islands have been undergoing retreat, due in part to dams on rivers and jetties along the coast strangling the flow of the sediment that historically replenished them. A future of threats to both people and wildlife lies ahead.

The coast of Texas may lack the rocky grandeur and crashing drama of the coasts of California or New England or the broad daily sweep of the ocean on the Atlantic Seaboard. Instead, a restless serenity pervades it, borne of the gentle tides, the softly churning surf, the everlasting wind over the low-lying landscape, the ebb and flow of its bird life. From the stories of the birds in this book, we may better realize not only that the future of our bird life is bound to our own but also that our stewardship reaches far beyond the bays, prairies, and woodlands of the Texas Coast.

LINDA M. FELTNER

CHAPTER 1

Drums of the Prairie

The Life and Hard Times of
Mr. Attwater's Chicken

THE STORY OF ATTWATER'S Prairie Chicken recounts a journey from a time of abundance to its status today as one of the world's most endangered animals. It is the story of warnings sounded early and long ignored, of research and action long delayed, of exploitation and indifference fought by scientists and citizens who wanted nothing more than to prevent the ultimate fate that this unique species, or any species, could endure. At the beginning of the twentieth century, this secretive, social grouse of the Texas coastal prairie pecked and strutted through the grass in vast numbers.

To the early settlers and explorers, the courtship call of the prairie chicken — the mournful warble that floated at daybreak over the damp grass in springtime — was the music of the prairie. In the words of one of the early wildlife scientists to study the bird and its habits, the prairie chicken was an intimate part of the "colorful and eventful early days in Texas. The prairie hen summons memories; it prompts old timers to recall when the range was free of wire fences and oil derricks, and rich grasses grew waist high." The bird's population today (ranging between several dozen and several hundred, depending on the time of year) survives only by the sustained intervention of wildlife and range managers, private landowners, captive breeders, and the federal government.

The prairie chicken, at first glance, does not look like a bird that would inspire artists and poets, much less like a bird that one day would have a national wildlife refuge devoted exclusively to its preservation. It is a light brown, darkly striped wild grouse that skulks through grass, eating insects, seeds, and bits of plants. The prairie chicken doesn't wander far from where it was hatched, spending its life hiding from danger, feeding, laying eggs in secluded grassy nests, and raising young. It flies for short distances, with bursts of powerful wing beats. The bird does not have a long life span, perhaps no more than a year or two, before owls or coyotes or stormy weather kill it. The seasons of its life cannot match the majestic flight of the Whooping Crane, the great, raucous clouds of waterfowl, the menacing verve of the raptors, or the epic migrations of shorebirds. It is a creature of the prairie grass, shy, secretive, elusive.

When it was first discovered, the Texas coastal grouse was standing in the way of a society emerging from its rural past into the modern world. The story of its decline follows the chronicle of not just the rise of modern Texas but also the disappearing prairie, from the time when great pastures of bluestem, switchgrass, and Indian grass lay among the rivers and woodlands, before the arrival of the plow and the pump jack. The bird's way of life, its vulnerabilities, its unyielding need for the prairie grass, would not enable it to withstand the assaults that were to come.

In the spring of 1893, an Englishman and former beekeeper with the improbable name of Henry Philemon Attwater went grouse hunting in the prairie grass of Refugio County, Texas. He shot two adult males there on March 27, and several weeks later he killed an adult female and three chicks in Aransas County, to the south, near the Refugio County line. In November, Attwater shot two more adult females in Aransas County and, in January of the following year, he headed east to Jefferson County, killing a male and female.

At the time, there was nothing remarkable about Attwater's hunting expedition. This Texas bird, along with other grouse species throughout the United States and Canada, was a popular and abundant game bird. The Greater Prairie Chicken, the Ruffed Grouse, the Lesser Prairie Chicken, and others ranged widely over the prairies, forests, and mountains of North America. Often slaughtered for sport and left to rot in piles, the Texas grouse inhabited the Gulf coastal prairie from southwestern Louisiana to the Rio Grande. Many accounts by early Texas explorers speak of encounters with "prairie fowl." German naturalist Ferdinand von Roemer wrote of seeing prairie chickens near Stephen F. Austin's headquarters at San Felipe; John Charles Beales, of the Rio Grande Colony, observed great flocks of them near Copano Bay, in 1834.

Attwater most likely did not realize it at the time, but he and the striped, chicken-like birds that were soon to be named after him were about to enter the annals of American ornithology. Today these specimens of Attwater's Prairie Chicken reside on Constitution Avenue in Washington, DC, on a broad wooden tray in case number M-16A, on the sixth floor of the Smithsonian Institution's National Museum of Natural History.

But on that spring day, Attwater, a self-taught amateur naturalist who had moved to Texas only a few years earlier, had set out in pursuit of science, not food or sport. He was collecting specimens for Major Charles Emil Bendire, one of the renowned ornithologists of the day. The nineteenth century would not consider Bendire at all unusual, but he would be difficult to imagine today: He was an Army officer who spent his spare time on post studying natural history and collecting biological specimens. Scores of career Army officers of that era, many of them in the medical corps, studied the wildlife of the regions where they were assigned, often making important contributions to scientific literature and collections.

The German-born Bendire spent much of his military career in the West and Pacific Northwest, studying birdlife and collecting eggs. He wrote *Life Histories of North American Birds* and corresponded with some of the leading naturalists of the time: Joel Allen, Spencer Baird, Thomas Brewer. He also served as honorary curator of the egg collection at the United States National Museum, as the predecessor of the Smithsonian Institution was then known, having donated his extensive collection to it. Bendire's Thrasher (*Toxostoma bendirei*) bears his name.

Field scientists throughout the country considered it an honor to collect specimens for him.

Bendire had been studying the geographic ranges of two closely related species of prairie grouse, or "hens," as they were then called, that were widely distributed throughout the grasslands of the American Great Plains: *Tympanuchus americanus*, the prairie hen, and *Tympanuchus pallidicinctus*, the lesser prairie hen. Bendire had received word from friends in the Army that a population of hens inhabited the grassy coastal plains of Texas. Determined to identify this unknown species, he requested specimens from colleagues in the field, Attwater among them. Attwater sent him the birds that he collected in Refugio and Aransas counties in March and April 1893.

Bendire studied Attwater's initial specimens and then announced in *Forest and Stream* in May of that year that he had identified a new species, which he named the southern prairie hen. He observed that the new species was "similar to *T. americanus*" but slightly smaller than that species and had less feathering on the legs. Granting it the status of a full species, he named it *Tympanuchus attwateri*, after Attwater.

Bendire, however, changed his mind. When Attwater sent him the specimens that he had collected later that year, he was forced to reconsider his original identification. "Since my preliminary description of this bird," Bendire wrote in *The Auk* in April 1894, "I have examined considerable additional material and am now compelled to consider it as only a well-marked race of *T. americanus*." The physical differences between Attwater's hen and the prairie hen simply were not great enough, in Bendire's reappraisal, to merit designation as a full species Thus, the southern prairie hen was only a subspecies, a geographic variant of the bird known today as the Greater Prairie Chicken. This taxonomic demotion was to have adverse consequences for the bird seven decades later, when federal conservation efforts began.

Bendire graciously acknowledged Attwater's contribution. "All the material received was kindly procured by Mr. H. P. Attwater of Rockport, Aransas Co., Texas," he wrote, "and generously donated by him to the U.S. National Museum Collection." Continuing his tribute,

Bendire named the bird *Tympanuchus americanus attwateri* "as a slight recognition for his trouble in obtaining these specimens."

In 1931, the American Ornithologists' Union renamed the prairie hen as the Greater Prairie Chicken, with a new scientific name, *T. cupido*; this species is designated today as *T. cupido pinnatus*, after the male's ear tufts, or pinnae. *Tympanuchus cupido attwateri* is the bird that will forever bear the name of Bendire's diligent collector: Attwater's Prairie Chicken.

Henry Attwater spent his remaining years advocating for Texas agriculture and horticulture and in wildlife study and conservation. He was among the first to promote the economic value of bird life to farmers, stockmen, and fruit growers. He lectured at county fairs, wrote books and newspaper articles, worked on behalf of legislation protecting wildlife (such as the 1903 Model Game Law, one of the nation's first legal protections for birds), published in scholarly journals, and, always, studied the birds and mammals of Texas.

But among his many pursuits, birds were his passion. H. C. Oberholser, author of *The Bird Life of Texas*, wrote in Attwater's obituary, "perhaps no one in Texas has done more to advance the cause of ornithology in the State than has Henry Philemon Attwater." Attwater died in Houston on September 25, 1931. He is buried in Hollywood Cemetery, in Houston, next to his wife Lucy Mary Attwater, a simple plaque marking his pine tree-shaded grave.

Scientists have studied the ways of the world's nineteen species of grouse to an extent rarely matched for other families of birds. This importance to researchers comes as much from the bird's role in society as from its biology and ecology. Grouse, such as the Red Grouse of Scotland, provide hunters several of their most prized game birds. They possess a wide range of mating behaviors that make them an ideal subject for the study of sexual selection, evolution, and sociobiology. Moreover, because of the birds' need for spacious, narrowly defined habitats, grouse researchers have established much of the foundation of our understanding of wildlife-habitat relationships and landscape ecology. Each species of grouse has adapted to a specific community of plant life, a natural assembly of different plants that feeds it and provides cover from its enemies, nests for its eggs, and shelter for it and its young.

At the time that Attwater was collecting his specimens along the Texas Coast, North America's twelve species of grouse occupied much of the continent's arctic and temperate regions north of Mexico, except for portions of the desert southwest and the forests of the American South. Grouse occupy coniferous forests from Alaska to Labrador and throughout the Cascade Range of the Pacific Northwest. They inhabit the hardwood forests of the Eastern Seaboard and regions of sagebrush from Montana to Nevada and Oregon. Ptarmigan inhabit the tundra and northern mountain ranges above the timberline. The prairie grouse — Greater Prairie Chicken, Lesser Prairie Chicken, Sharptailed Grouse — range throughout various regions of the Great Plains, the great expanse of grasslands that spread across the interior of North America.

All extant grouse throughout the world have declined in population, some severely, others to a lesser degree, as their historic natural habitats have disappeared. The prairie grouse, because of their intimate dependence on grasslands, have suffered the worst loss in population numbers and extent of habitat and face the greatest need for recovery efforts. But none of the extant grouse species has been pushed so near the precipice of extinction as has Attwater's Prairie Chicken. A close relative, the Heath Hen (*T. cupido cupido*), offers a cautionary tale.

This eastern race of the Greater Prairie Chicken once flourished in the grassy oak woodlands and pine barrens of the eastern US, from Maine to Virginia. Like the Texas coastal grouse, it was for generations an abundant game bird. By the late 1920s, the population had dwindled to thirteen individuals, mostly males, all on Martha's Vineyard, in Massachusetts. By 1928 only one bird, a male, was known to exist. In 1930 the journal *American Game*, the bulletin of the American Game Protective Association, published an article entitled "One Lone Heath Hen Still Survives," with a photograph of the bird standing on a rock. After spending the summer in seclusion among the dense scrub oaks, the bird would reappear each spring near the farm of Mr. James Green. Mr. Green would look after the bird and assist visitors who wanted to see him. The bird wandered around his traditional booming ground — the open area where prairie chicken males perform their courtship ritual — that lay along the highway. No female ever appeared. This went on for several years. After a sighting on March 11, 1932, the bird was never seen again.

The ancestral home of Attwater's Prairie Chicken lay within the coastal tallgrass prairie that extended from Opelousas and Bayou Teche in southwestern Louisiana, running along the upper Texas Coast, before turning at the Coastal Bend and fading in the dry heat of South Texas, near the Rio Grande. This prairie is the southernmost grassland within a succession of plant communities that lay across east-central Texas. At about 600 miles long and up to sixty miles wide in some areas, the Texas coastal prairie is also the southern boundary of the tallgrass prairie of the Great Plains. The North American plains and prairies are part of the vast temperate grasslands that span the interiors of continents worldwide, such as the pampas of Argentina and the steppes of Russia. These regions share the melancholy distinction of being among the world's most endangered biomes, for their soils, enriched by deep-rooting grasses, have led to their conversion to agriculture on an immense scale. As befits the geography of its home range, Attwater's Prairie Chicken is the southernmost-occurring grouse in the world.

The Texas coastal prairie possessed, in the words of Mary Austin Holley, one of Texas's early chroniclers, a "peculiar beauty," with its woodlands and prairie grass. Unlike the popular image of the American prairie as a vast treeless expanse of unbroken, undulating grasslands, the Texas prairie was a tapestry of large pastures, dotted with mottes of oak and laced with rivers and streams that were fringed with dense gallery forests of willow and sycamore.

Upland forests, with pines and oaks east of the Brazos River and oaks to the south, rimmed the crescent of the Gulf Coast almost continuously; in many areas the timberlands reached close to the shore. Much of the prairie is flatland that often gives way to broad rises and falls, like the sea under a gentle wind. Over the palette of tan and pale green grass, forbs — non-woody, non-grass species such as wildflowers — dusted the pastures with a galaxy of petals of red, blue, orange, and yellow.

The coastal prairie is a landscape of mud, sand, and silt laid down by the advance and retreat of the Ice Age Gulf and the meanderings of ancient rivers, the forebears of our modern waterways. The native prairie that these natural forces had sculpted was marked by a variety of soil types, topography, and drainage features that gave rise to communities of grass that differed in species, height, and density and, ultimately, in their value to the prairie chicken. Even slight changes in soil and elevation could profoundly alter a pasture's composition. One area might be marked by spiny aster or sumpweed, while others nearby will have dropseed or Texas grama or the climax grasses such as bluestem and Indian grass.

The profile of grasses — some tall as the flanks of a bison, some close to the ground, some thinly arrayed, with others in thick bunches — reflects not only their differing species and soil but also the other creative forces that mold the prairie's ever-changing face: rainfall, grazing, fire, and drought. The prairie would change, from year to year and season to season; a nurturing pasture one year might turn into a drought-scorched, barren plot the next. As wildlife scientists were to discover, much of the coastal prairie was not suitable for Attwater's Prairie Chicken; only grasslands of a very specific character would serve the bird for all its needs for food, shelter, and raising young.

The natural creative forces of fire and grazing allowed the prairie to remain a diversified community of grasses, a community of different species and at differing stages of growth that produced the special character needed by the prairie chicken. A prairie that was never burned or grazed became dense and rank, unusable by grassland birds; eventually the woody invasive trees such as running live oak would create their own community and the prairie grass, and the birds that depended upon it, would disappear. Fire cleaned out dense, dead debris and undergrowth, returned nutrients to the soil, and allowed a new generation of grass to emerge. Grazing, first by bison and later by domestic livestock, helped maintain the vital diversity of density in cover.

At the turn of the twentieth century, an estimated 6.5 million acres of native prairie occupied portions of the Texas coastal plain counties between the Sabine and Nueces Rivers. The most reliable historic records of the bird's occurrence come from this region, with the largest and most contiguous populations occurring between Galveston Bay and Copano Bay. In the South Texas desert savannah, grasslands were often sparse due to drought and they probably did not support significant numbers of the grouse. Records of sightings there were infrequent and of less historic validity. Nevertheless, prairie chickens are believed to have occurred at least occasionally in small numbers within the state's far southern counties.

The growing threats to Attwater's Prairie Chicken slowly revealed themselves in the early years of the twentieth century. Hunting, as it was historically pursued, was soon recognized as the most immediate danger to the bird's survival. As early as 1916, Henry Attwater himself, in an address to the Scientific Society of San Antonio entitled "The Disappearance of Wild Life," warned of the prairie chicken's decline and called for an end to the open hunting season. But Attwater was also speaking out against the unchecked warfare on all birds — songbirds, shorebirds, doves, hawks, egrets, and others — that were believed to be either enemies of commerce or an economic resource that could be harvested without restraint.

The prevailing belief held that the vulture transmitted anthrax to cattle; that the pelican threatened the fishing industry; that doves damaged wheat and oat crops. Whatever game laws were in place at the time were largely ignored or were being rescinded, Attwater said, with the resulting disappearance of wild life. Furthermore, he insisted, it was time for us to recognize the economic benefit of birds for their control of pests and predators. Attwater concluded, with a tinge of British understatement and possibly his eponymous grouse in mind: "[I]t will be well to remember that there is no recovery for an extinct species." The hostile attitude toward wildlife persisted for decades; a 1954 issue of *Texas Game and Fish*, the predecessor of *Texas Parks & Wildlife*, contained an article entitled "Let's Declare War on the Crow."

Despite early attempts at restriction of hunting, enforcement was practically nonexistent. Moreover, over half of Texas counties claimed exemptions from all hunting and fishing regulations in the decades around the turn of the century. The state's original hunting regulatory agency, the office of the Game, Fish, and Oyster Commissioner, was instituted in 1907, but funding was scant and the agency was poorly staffed. The practical consequence for the prairie chicken was that hunters killed them at will.

In the early 1900s, wagon train parties would move into the Bernard Prairie, the rich prairie chicken territory along the San Bernard River, in Colorado County, and set up shooting camps. People were assigned to keep count of the number of shots that were fired versus the number of birds that were killed. Birds were killed by the thousands. One camp account tells of a hunt yielding ten piles of birds, each with approximately 150 birds. The birds were so numerous in some areas that poultry growers had to feed their domestic fowl under the house to prevent the prairie chickens from eating the grain. The prairie chicken was viewed as a popular and seemingly inexhaustible game bird throughout its range. The birds provided food for cattle camps or were simply shot for sport. Shooting teams would compete for the most birds killed; the team with the smallest tally would have to pay the expenses of the hunting trip.

In 1925, the Texas Legislature enacted the Boyd-Hubby Game Bill, a comprehensive law that extended hunting restrictions to much of native game wildlife. The statute was blunt in its assessment of the problem: "The fact that there are now no adequate laws for the preservation, propagation, and protection of the wild game animals and birds of the State, which are rapidly disappearing . . . creates an emergency and an imperative public necessity . . . " Under this bill the Lesser and Attwater's Prairie Chicken could be hunted for only ten days in September and with a daily limit of five birds and a total limit of ten. The Greater Prairie Chicken had already disappeared around 1920 from its Northeast Texas range. After a sighting near Vinton in 1919, Attwater's Prairie Chicken was never again reported in Louisiana.

When Texas governor Dan Moody visited the Eagle Lake area for a prairie chicken hunt in 1927, the Eagle Lake Headlight reported the event with the large bold headline: "GOV. DAN MOODY HERE FOR HUNT." The governor was given the honor of killing the first three birds. As he proudly held up one of them, he said that this was not only the first prairie chicken he had ever killed but also the first that he had ever seen. Turner E. Hubby, the State Game, Fish, and Oyster Commissioner after whom the bill was named, had accompanied the governor on his Eagle Lake trip.

Complete protection from hunting finally came with the enactment of House Bill 30, in 1937, during the 45th Legislature of the Texas House of Representatives. The bill read: "An Act declaring it unlawful to take, hunt, trap, shoot or kill any prairie chicken in Colorado and Austin Counties, Texas . . . " The bill was immediately amended, with the reference to Colorado and Austin counties deleted and replaced with "the State of Texas." HB 30 passed 97 to 6 and was to be in effect for five years. Subsequent legislatures renewed the ban. After decades of relentless exploitation, Attwater's Prairie Chicken could never again be legally killed in Texas.

As important as the hunting ban was for the bird's survival, it did not stop the greatest threat of all: the ever-increasing loss of native prairie. Rice farming in Colorado and Austin counties, cotton and sorghum farming in Refugio, Goliad, and Aransas counties, and oil and gas exploration and urban development in Harris, Fort Bend, Galveston, Jefferson, and Chambers counties were to plague the fortunes of the prairie chicken throughout the century and beyond.

The first public champion of Attwater's Prairie Chicken, the person who conducted the first scientific study of the bird's life history and its retreat from its historic range, was a wildlife scientist and conservationist by the name of Valgene W. Lehmann. A native Texan born in Washington County in 1913, Lehmann worked for various state and federal wildlife agencies, eventually becoming the first staff wildlife biologist for the King Ranch in 1945. No one person did more for the Attwater's Prairie Chicken than Valgene Lehmann.

Tall, slender, weathered from years of working in the outdoors, Lehmann made little attempt at diplomacy in dealing with associates whom he felt were not doing their job adequately. Former colleagues describe him as likeable but demanding and often difficult in his fierce devotion to the welfare of the prairie chicken. One incident in later years involved Lehmann conducting a helicopter survey over some pastures where cattle were grazing under a prairie chicken management program. Lehmann discovered that the rancher was running too many head of cattle; he had the pilot land the helicopter, stormed up to the rancher's house, pounded on the door and roundly chastised him for the infraction. The fact that the rancher in question was one of the few private landowners at the time devoted to prairie chicken conservation did not deter Lehmann from his disciplinary mission.

At the request of Walter P. Taylor, head of the Texas Cooperative Wildlife Research Unit at College Station, Lehmann assisted in the development of a plan to conduct a census of the coastal grouse's population and to delineate the extent of its remaining habitat. The decades of gathering concern had finally led to recognition of the need for a systematic investigation of the prairie chicken and its environment.

In 1937–1938, Lehmann, assisted by officials of the U.S. Fish and Wildlife Service, the Texas Game, Fish, and Oyster Commission, faculty of the Agricultural and Mechanical College of Texas, state game wardens, and numerous landowners, visited approximately 90 percent of all farms and ranches that were thought to support prairie chickens. Lehmann's work resulted in the seminal 1941 monograph *Attwater's Prairie Chicken, Its Life History and Management*; it described the status of the bird in stark and unsettling detail.

The bird's population had plummeted from its historic levels. The survey concluded that throughout the coastal grasslands, only about 8,700 prairie chickens remained. Estimates of previous numbers, while difficult to make with much accuracy, traditionally ranged from half a million to upwards of one million birds based on the amount of suitable habitat and the density at which prairie chickens were known to occupy grasslands.

Attwater's Prairie Chicken appeared to inhabit the Texas coastal prairie at less than one percent of its population at the beginning of the twentieth century. About half of the surviving population — an estimated 4,200 birds — could be found around the Coastal Bend. The bird had declined or disappeared in twenty-five of the twenty-seven counties, from Orange County in the east to Cameron County in the south, that once provided a home for them.

The loss of prairie grass was equally grim and, indeed, lay at the heart of the bird's decline. Of the original estimated 6.5 million acres of prairie chicken range, fewer than 458,000 acres remained, representing a loss of 93 percent. This surviving area was, at best, sparsely populated with chickens, with the birds occupying only 30 percent or less of any apparently suitable grassland. Only four counties — Refugio, Colorado, Brazoria, and Austin — had sufficient remaining habitat where it was believed the bird had at least a chance for stable populations.

Lehmann described the bird's life history: its diet and feeding habits, movements, courtship, mating, nesting, and brood raising. The enduring value of his work, however, lay in the recognition that Attwater's Prairie Chicken had precise needs for survival that could be provided only by a community of prairie grass that had specific characteristics of species, composition, and density. Different stages of the bird's life history each needed a different type of prairie grass assemblage.

Lehmann observed that the grasslands that supported the most robust populations of chickens were well drained, characterized by slight ridges and knolls, and with a diversity of plant species — grasses, wildflowers, weeds, sedges, rushes, legumes — distributed over areas that varied in density and height. Chickens require grassy areas that range from light to moderate to heavy cover, with some areas characterized by bunched grasses surrounded by less dense, shorter grass. A slightly rolling topography also contributes to the optimal chicken pasture for the birds to have temporary sanctuary on the top of knolls and ridges during heavy rains and floods. Surface water is important primarily in drought; sufficient water is usually obtained from the diet. Overgrazed, scalped pastures and dense, overgrown pastures were of no value to the chicken.

The grasses provide shelter for roosting, nesting, and hiding from predators; insects nurtured by grasses and forbs provide the predominant food component for chicks while the adults eat mostly seeds, flowers, and leaves of weeds and wildflowers, such as coreopsis, buttercup, and blue-eyed grass. Chicks will eat plant matter and adults will eat insects as a smaller fraction of their diet, depending upon availability and time of year. Insects, for example, comprise a smaller portion of the diet in winter.

The most important feature of a chicken pasture is "clumped midgrass," characterized as moderately dense stands of knee-high grasses, such as bluestem or Indian grass, that are surrounded by expanses of shorter, less dense cover. These grassy islands provide nesting habitat and shelter from predators; birds have been known to nest in more open and vulnerable areas, but the clumped midgrass is the location most preferred by a nesting hen.

The open areas surrounding the nest site allow the chicks to move about and feed; the female adult tends to the brood but the chicks feed themselves on insects — beetles, grasshoppers, caterpillars, spiders, leafhoppers — soon after hatching. Grass communities with rich spreads of forbs are important, especially near areas where the chicks hatch, since these plant species support abundant populations of insects.

Finally, the prairie chicken needs a "booming ground," areas of short cover ranging up to several acres in size where the males, each spring, engage in the courtship dance for which Attwater's Prairie Chicken is renowned; Lehmann described the courtship antics as "unbelievably weird." Males will also use man-made areas for booming, such as oil wellheads, gravel roads, pipeline rights-of-way, and under high-voltage power lines, as well as areas of naturally occurring short grass or hardpan flats. Some booming grounds were used year after year; when those historic grounds were destroyed, the chicken population was at risk of severe decline or eventually dying out.

The booming ground is the communal area where the chicken's life history begins and where the males display the extravagant courtship dance. This area, known as a lek, has been described as "the social center of prairie chicken ecology," since most of the later stages of the bird's life history — nesting, brood raising, feeding, roosting, seeking shelter — take place within approximately a mile of the booming ground.

Beginning in late February–early March, males gather on the booming ground in the mornings and late afternoons to establish territories and to attract hens. They will confront one another and begin the courtship display of spreading their wings, stretching the head and neck parallel to the ground, rapidly stamping the ground, elevating the ear tufts, and puffing out their golden-colored neck sacs while they emit the "booming" call. This distinctive call — a throaty, low-pitched trill — could travel a mile across the prairie on quiet days. Fights sometimes occur, with two males jumping in the air and flapping their wings at each other, feathers flying, like a pair of battling barnyard roosters.

Females are attracted to the lek by these courtship displays and eventually mate, usually with one of the dominant males. The hen then departs the lek and seeks out the clumped midgrass to build a nest and lay her dozen or so eggs. These nests are well concealed; range management personnel speak of being right next to a nest without seeing it. After a four-week incubation period, the eggs hatch, one per day, and the hen leaves the nest with the brood for the more open cover surrounding the clumped midgrass.

For about the first three weeks of life the chicks follow the mothering hen to feed throughout this open area, usually within a half mile of the nest site. Chicks and adults use cow trails, when present, for ease of movement. During this early period, the hen frequently broods the chicks — having them huddle beneath her extended wings — because the young cannot maintain their own body temperature. As the chicks grow, the frequency of brooding lessens. The chicks grow and develop rapidly, a necessity for their entrance into the hazardous world of the prairie; by the age of two weeks they can fly short distances.

The brood eventually breaks up around six to eight weeks after hatching, although in some cases chicks stay with the hen until later in the year. By twelve weeks of age they are full-sized and can begin their independent lives as adult birds. A typical day for adult birds in non-breeding season will see them feeding at sunrise, roosting during the day, and feeding again shortly before sunset.

As summer approaches, the flock will seek out shade under shrubs and taller grasses, away from the feeding areas where they spent the first several weeks of life. In winter the flocks will use areas of moderate to heavy cover for shelter from weather and predators, such as the raptors that migrate over the Texas Coast each fall. Here they stay for several months, gathered into gender groups, after which the annual return to the booming grounds begins in the spring, renewing the cycle of courtship, breeding, and nesting.

In addition to describing the bird's basic life history, Lehmann also provided a crucial insight: Since so little of the native prairie remained, mostly under private ownership, the only way for suitable prairie habitat to be sustained was through active stewardship by landowners. He called for moderate cattle grazing, judicious use of fire and mowing outside the breeding season, brush control, and other measures to maintain grassland characteristics suitable to the prairie chicken. Lehmann applauded the policy of the Refugio County landowner who leased his pasture to an oil company: He had it written into the contract that if a worker so much as showed up on site with a gun, the lease was immediately terminated.

Throughout the bird's original range, the coastal prairie varied widely in quality. Many areas were poorly drained or lacked sufficient variety of plant species or densities. Lehmann had estimated that ideal conditions for the prairie chicken probably did not exceed about 15 percent of the original native habitat. These areas were thought to support a population of chickens at a ratio of approximately one bird per acre. As conditions declined, some pastures could still support chickens but at a greatly reduced capacity — an estimated one bird per ten to fifty acres.

Lehmann's study caught the attention of the Department of the Interior, which issued a national news release in 1940 about the bird's perilous state. "The Attwater's prairie chicken . . . is rapidly disappearing," the agency warned, "and will soon join the ranks of . . . extinct birds unless adequate measures are taken soon." Ecologists and wildlife scientists in later years would greatly enhance knowledge of the bird's complex way of life. Lehmann's study presented the bird's life history in broad terms; it would be several decades before the bird's reproductive ecology and habitat requirements were better understood. But Lehmann's pioneering work was more than a pivotal scientific study; it was a portent, and a call for action.

Attwater's Prairie Chicken is "on the verge of extinction," Lehmann said in a 1939 address to the Texas Academy of Science. "The boom of the Attwater's prairie chicken . . . [is] becoming increasingly faint . . . unless man acts quickly and forcefully in the prairie chicken's behalf . . . . Save portions of their homeland, and the boom of the Attwater's Prairie Chicken, drums of the prairie, may beat forever. Remain inactive, and they surely die." Somewhat ominously, Lehmann once referred to Attwater's Prairie Chicken as "the heath-hen of the South."

A diverse array of assaults, both natural and man-made, had led to the population collapse. Storms and floods were well documented for their toll of hundreds of birds, especially in Colorado and Jefferson counties. A 1917 storm in Jefferson County, previously an area of abundant chickens, resulted in scarcity of the bird there for several decades. Lehmann's notes describing the aftermath of a sudden and heavy rain in Colorado County draw a poignant scene:

> The prairie has been transformed into a miniature ocean dotted by tiny islands that previously had been the tops of knolls and ridges. On these islands sit wet and bedraggled prairie chickens . . . that seem confused and astounded as I by the sudden change in their environment . . . . Problems due to hawks, skunks, and other predators seem so petty when excessive rain destroys virtually everything at a single stroke.

Heavy rains during the crucial nesting and brooding period of March–June would severely reduce the yield of new chicks. Encroachment of woody vegetation, such as mesquite, live oak, and McCartney's wild rose, resulting from suppression of prairie fire occasioned a substantial loss of suitable habitat. Drought devastated the bird on two fronts. First, grass cover dies, depriving the chicken of nesting habitat and protective shelter, rendering it more vulnerable to predators. Without sufficient protective cover, predation becomes especially severe in the fall, during raptor migration. Second, forbs die out, reducing plant material and insects as food sources.

But Lehmann was unsparing in his indictment of the ultimate reason for the bird's decline, regardless of agency: the "spread of civilization," as he grandly states. After all, had not Attwater's Prairie Chicken maintained stable and abundant numbers for millennia, even with their storms and droughts and predators? In Lehmann's view, the effects of these natural events were intensified by man's destruction of the prairie lands.

Pastures were burned wholesale to enhance forage for livestock grazing and were mown for hay; these denuded areas were useless to the prairie chicken. Unregulated mowing and burning, especially during the nesting season, resulted in destruction of nests and broods and in grasslands that lacked the necessary variety of species and density. Conversely, prairie fires were discouraged by governmental land policy in the 1940s, leading to intensified loss of prairie to invasive species. The spread of oil derricks, monuments to the emerging oil industry, often stood in the heart of prime territory, forcing out the prairie chicken and other grassland birds.

Changes in traditional cattle management led to overgrazing, which removed grass cover and aided the spread of trees and bushes. Predation by hawks, skunks, coyotes, and feral cats killed adults and chicks; skunks, raccoons, and snakes ate eggs and destroyed nests. As habitat disappeared, birds would crowd into smaller areas, making them more susceptible to floods, predation, starvation, and nesting failure. But nothing was more permanent in its damage, more vast in its destruction of the home of the prairie chicken, than the Gulf Coast rice industry.

The origin of rice farming in Louisiana and Texas embodied the themes of the nineteenth-century American saga: the coming of the railroad, immigrants seeking a better life, developers promoting cheap land, modern technologies replacing traditional methods, the decline of one economy and the rise of another. Rice farming in the Gulf Coast was, in the words of one historian, "one of the last agricultural frontiers of America." From its beginnings in southwestern Louisiana, it spread rapidly throughout Southeast Texas. Within only a few decades, rice agriculture had become a powerful and pervasive economic force, laying the foundation for much of the region's early wealth.

The coming of the railroad in the 1880s opened South Louisiana to a wider world. Land developers soon embarked on an aggressive promotional campaign to entice settlers, especially farmers from the Midwest, to the region. They had been burdened for decades by high interest rates, severe weather, and grasshopper infestation and were primed for the exodus to the cheap land of South Louisiana. The farmers quickly realized that their traditional crops of wheat and corn were not suited to the region's cycle of heavy rain and droughts so they learned the new methods demanded by the complexities of rice agriculture. Rice farming became their new way of life.

From its beginnings in Calcasieu and Cameron parishes in Louisiana, commercial, mechanized rice agriculture soon followed in Texas, in Orange and Jefferson counties. County by county, decade after decade, rice production in Texas grew steadily westward, consuming vast tracts of the coastal prairie, at rates approaching 200,000 acres a year. Rice acreage grew from 2,000 acres to 264,000 acres between

1895 and 1910. By the late 1930s, two million acres of coastal prairie had been plowed under for agriculture. More than a million acres were lost to rice farming alone. The growth of rice farming would continue for decades.

Colorado County, some of the richest chicken territory in the bird's historic range, was hit especially hard. Lehmann had identified six prime chicken areas there; one by one, he saw them vanish, negotiated by the "rice barons," as he called them, to be converted to rice fields. In Colorado County alone some 84,000 acres of prairie chicken habitat had been destroyed. Lehmann was unsparing in his scorn: "The prairie chicken is doomed [in the county]," he wrote, "just as if we were to go out and kill the last one." The prairie chicken, its grassy home, its welfare, and its survival did not enter the thinking of the businessmen and farmers who created the rice industry. The bird was simply not a part of the scheme of things. If one of the tawny, striped birds were chased away or a gaggle of fluffy yellow chicks scattered or an unseen nest crushed by a tractor, there were probably plenty of birds and nests in the next county.

The years following Lehmann's original census witnessed a continued decline in population. Lehmann and his colleagues conducted three more surveys, each one showing a severe reduction from the time before. The great storms that struck Texas in the 1960s had also exacted a heavy toll. Hurricane Carla, the most intense storm of the century to hit the Texas Coast by that time, made landfall over the eastern tip of Matagorda Island in 1961, wiping out almost all the remaining chickens in Chambers, Jefferson, and Galveston counties. Hurricane Beulah, moving ashore near Brownsville in 1967 with 28-inch rains, killed over a thousand birds in Refugio, Goliad, and Aransas counties. Prairie continued to disappear under the plow. The growing urban areas of Harris and Galveston counties added to the loss. By the end of 1967 scarcely more than a thousand birds remained in the wild. Clearly, by the late 1960s, time was running out on Mr. Attwater's chicken.

As the 1970s approached, the bird's prospects brightened, as universities, scientists, wildlife managers, conservation groups, and government agencies began to take action on the bird's behalf. Texas A&M University and the Texas Parks and Wildlife Department instituted research programs on the prairie chicken, with twenty-six birds being relocated from Ellington Air Force Base, near Houston, to College Station for study. Conferences were held, with panel discussions on prairie chicken ecology, captive breeding, and habitat management; over the next few years academic research on the prairie chicken's ecology began, much of it under the direction of Dr. Nova J. Silvy of Texas A&M University.

Several early federal laws identifying and offering minimal protection to endangered species became law; Attwater's Prairie Chicken was officially listed as endangered in 1967. The Endangered Species Act, the flagship statute for protection of species at risk of extinction, was enacted in 1973. Unfortunately, the bird's status as a subspecies would not allow the highest level of protection afforded under the law to birds of full species status. Finally, however, after decades of loss and retreat, it appeared that the wild population of the Texas coastal grouse had not only a chance for survival but also perhaps for recovery.

One morning in March 1965, a group of local landowners, bankers, sportsmen, wildlife scientists, and leaders of several national conservation organizations met in a damp pasture in Colorado County. They had come together, at the urging of Valgene Lehmann, to discuss how they might finally create a refuge for the prairie chicken. They were gathered at a remnant of surviving tallgrass prairie that straddled Coushatta Creek, along the west bank of the San Bernard River. The owners — I. V. Duncan and David Wintermann — had generously agreed to sell it at $100 an acre, half the market price. This property was some of the last native prairie chicken grassland in the county, and the concerned citizens wanted to keep it in perpetuity for the endangered bird. The land's first caretaker was Thomas T. Waddell, a friend and colleague of Lehmann, who was widely known for his dedication to educating local farmers and ranchers about preserving wildlife — especially the prairie chicken — through wise land stewardship and prairie conservation. The sanctuary that Henry Attwater had pleaded for sixty years earlier had finally become a reality.

The U.S. Fish and Wildlife Service eventually took over the property, creating the Attwater Prairie Chicken National Wildlife Refuge on July 1, 1972. (A misspelling in the congressional paperwork setting up the refuge led to the incorrect term "Attwater" in the name.) Later acquisitions of additional parcels brought the refuge to its current size of 10,541 acres. The fundamental mission of the refuge was to provide a haven for the prairie chicken by maintaining in perpetuity a suitable parcel of native coastal prairie. In pursuing this mission, the refuge would serve as the foundation for all future conservation efforts, from ecological research and public outreach to release of captive-bred birds, prairie restoration, support to private landowners, and the development of range management methods, such as prescribed fire and controlled cattle grazing. In fact, both cattle grazing and fire would emerge as the primary land management tools at the refuge and on private lands with chicken populations.

By 1975, the statewide population of the prairie chicken had grown to 2,240 birds, twice as many since the 1967 census and the highest since the 1950s; that was to be the high-water mark of the modern era for the prairie chicken. Then, a slow, steady decline began that has persisted to the present day. Except for a few years in the mid-1980s when populations remained stable, the bird's population continued to decline over the next three decades under the grind of its historic nemesis: the loss of prairie, aggravated by seasons of extreme weather.

Year after year, the bird was to disappear from those areas that for generations had given the prairie chicken a home. By the early 1990s, the prairie chicken had retreated to a few redoubts, mostly in Refugio, Colorado, and Austin counties. The continued expansion of rice and row-crop agriculture throughout the chicken range; urban growth in Houston and Galveston; overgrazing in Goliad, Victoria, and Fort Bend counties; blackland farming in Refugio County; loss of native prairie to huisache, McCartney's wild rose, running live oak, Chinese tallow, and other invasive species: all contributed to the unabated loss of habitat and decline in population.

A poignant photograph in the archives of the prairie chicken refuge shows bulldozers in Galveston scraping the grass away from the site of a planned residential neighborhood. This area was the last major chicken pasture in Galveston County; the birds were rounded up and carted off to Victoria County. Very few survived the relocation. Within ten years, Fort Bend County lost its last 12,000 acres of native prairie to agriculture. Large ranches in Victoria County were subdivided and leased to absentee cattle ranchers and overgrazing became the norm. East Harris County, with its massive growth spurred by the Manned Spacecraft Center, lost huge areas of chicken range. U.S. Air Force Col. Donald Robison, former Commanding Officer of nearby Ellington Air Force Base and this author's uncle, recalled a lone male prairie chicken — an "ol' rooster," as he called him — that would strut around the runway, and then one day was never seen again.

During much of this period, portions of the chicken range endured cycles of extreme weather: several years of drought followed by several years of heavy rains during the critical nesting season, followed by years more of drought. Tropical storms — leaving record-breaking amounts of rain — hit the upper coast in 1979. Hurricane Allen struck South Texas in 1980. Heavy spring rains during the 1985 brood season slashed the statewide population by forty percent, to around 870 birds. The population crashed again by the mid-1990s, from around a thousand birds in 1988 to fewer than a hundred by 1995 — a ninety percent decline in an already stressed population. By the end of the decade, Attwater's Prairie Chicken had disappeared from all private lands and was found only on the Eagle Lake refuge and on a private prairie preserve in Texas City. The species had approached a new precipice of extinction.

At the approach of the new millennium, Attwater's Prairie Chicken stood at a paradoxical time in its history. Scientific knowledge of the bird had reached unprecedented levels; much research had been conducted on its life history, habitat requirements, the effects of fire and grazing on grasslands, captive breeding protocols, reproductive behavior, genetics, disease, parasitism, and land management techniques for maintaining the vital fabric of grass, wildflowers, and open space. A national wildlife refuge had been founded for its preservation and scores of dedicated professionals were working on its behalf. Private landowners of large tracts of native prairie had been enlisted in conservation efforts. The Coastal Prairie Conservation Initiative, a consortium of private landowners, governmental agencies, and scientists, was instituted to develop a program of restoration of native prairie within the prairie chicken's range. Most important of all, a captive breeding program was developed and began releasing birds to the wild. And yet the bird's population continued to decline.

One particularly disheartening moment came in 2005, on the prairie chicken refuge in Eagle Lake. It had been a good brood season that year, with eighty chicks surviving the first few perilous weeks of life. The refuge staff had been diligently monitoring them and saw they that were doing well. Then, on Memorial Day weekend, a heavy spring storm moved across the area, killing almost all of them. The severe drought of 2011 was followed by three years of heavy spring rains between 2014 and 2016, all destroying many nests and chicks. "We saw chicken eggs floating away," says former refuge manager Terry A. Rossignol, sadly recalling the infamous Tax Day Flood of 2016, when huge tracts of the refuge were underwater for days. Three thousand acres of the refuge were under water in that flood, killing most of the birds. The few survivors retreated to higher areas, but the predators went there as well. "We went from a twenty-year high to a twenty-year low in prairie chicken numbers," says current refuge manager John Magera.

It is an enduring irony that a major deterrent to the bird's recovery is essentially the purpose that this grouse serves in the natural scheme of things. Predation — death of birds by hawks, owls, and coyotes, destruction of nests and eggs by snakes, raccoons, and opossums — is simply the manifestation of the prairie chicken occupying its natural location in the food web, serving as a food source for animals higher in the grassland community.

Mortality is inherently high in prairie chicken populations. Statistics will vary depending upon local conditions, but on average only three nests out of ten successfully hatch chicks. Of the chicks that do hatch, four to five (out of an average of ten to twelve) will not live to see the age of eight weeks. This mortality results largely from one of several factors: failure to thrive from poor maternal nutrition, separation from the brood in dense vegetation, poor insect availability, heavy spring rains, or predation by hawks and owls. Then, upon reaching adulthood, the birds must contend yet again with their natural predators. Most prairie chickens do not live longer than two years.

The vast tracts of grasslands that historically supported widespread populations allowed the species to absorb and recover from these assaults. In fact, some ecologists believe that the prairie chicken in the pre-settlement era represented one large, loosely contiguous community of birds, a "meta-population"; if one region was devastated by, say, drought or a hurricane, the birds in adjacent areas could eventually re-populate the depleted area. Aransas and Refugio counties, for example, whose prairie chicken populations were reduced by about 75 percent as a result of flooding from Hurricane Carla, saw their populations recover substantially within about five years. Following the 1917 hurricane in Jefferson County, the chicken population there recovered after about fifteen years.

But as habitat became reduced and fragmented and the numbers of chickens became perilously low, the losses that were part of the prairie chicken's natural cycle of life took on a more threateningly permanent lethality. With small, isolated populations still having to deal with these natural forces, they are less likely to recover; the population must struggle against what Terry Rossignol calls a "vortex of extinction." By the end of the 1990s, the bird's plight was becoming desperate; scientists from the International Union for Conservation of Nature had forecast that, in the absence of vigorous intervention, wild populations of Attwater's Prairie Chicken would be extinct by the year 2000.

Despite the massive loss of native prairie, there still exists today enough pasturage of suitable habitat that could potentially support a naturally sustainable population. Most of this area — approximately 60,000 to 70,000 acres of native bluestem prairie — lies on private ranches within Goliad and Refugio counties. With almost 11,000 acres on the Eagle Lake refuge and 2,300 acres on the Texas City Prairie Preserve (established in 1995 by a donation of land by the Mobil Oil Company and managed by The Nature Conservancy), the prairie chicken has at least sufficient space to maintain minimally sustainable populations. For this to happen, however, population numbers need to reach a critical minimal level for the bird to overcome its natural mortality rates and achieve sustainability. Therein lies the major challenge facing Attwater's Prairie Chicken.

Wildlife scientists today have determined that two serious impediments stand between the prairie chicken and its long-term survival. One of these is nesting success, which is the percent of nests that yield chicks. The other is the survival rate of chicks hatched in the wild. The latter of these — chick survivability — is considered the more serious problem, because if chicks do not survive to become part of the following spring's breeding population, the population will continue its decline. Even with reduced habitat and the high rate of nesting failure, researchers estimate that if the chick survival rates could be improved to a sufficient level, the bird would have a chance at sustainable populations.

Lying at the very heart of the bird's future is the captive breeding program. This was first attempted in 1968, at the Poultry Science Department of Texas A&M University, when the university obtained thirteen males and thirteen females that were captured from the wild. The purpose of the captive breeding program is to provide chickens for repopulation into depleted areas, mainly the Eagle Lake refuge. The importance of captive breeding cannot be overstated; each prairie chicken alive in the wild today is a descendant from a captive-bred bird.

Captive flocks are managed at places such as the Houston Zoo and the Fossil Rim Wildlife Center, near Glen Rose. These flocks produce chicks that are then nurtured for six to eight weeks, at which time they are transferred to the refuge or other habitats. At the refuge they are first placed in acclimation pens, which are outdoor enclosures designed to allow the birds to become accustomed to the natural habitat while being protected from predators. After about two weeks they are released to their new life on the prairie. Release usually occurs in the summer, prior to hawk migration in the fall.

The captive breeding program has discovered its own set of difficulties, infertility of hens and failure to thrive of young chicks being among the most urgent problems. Hannah Bailey, Curator of Birds and Natural Encounters at the Houston Zoo and director of the breeding program there, describes the raising of captive prairie chickens as an "art"; the reasons for poor productivity in captive animals are largely unknown. The main challenge facing the breeding program is to increase its productivity; it is estimated that at least 400 to 500 chicks a year need to be released to the wild to overcome mortality rates. Although numbers of chicks have varied from year to year, on average the productivity of the program is approximately half of what it needs to be.

Mortality of the released chicks is high, at around 75 to 80 percent. The surviving released chicks breed the following spring and produce their own brood. It is the survivability of this generation of chicks — the wild-hatched chicks of released captive-raised birds — that presents one of the great challenges to the recovery of the species. Moreover, another intruder has appeared on the scene that is believed to be a major factor suppressing chick survivability: the red imported fire ant.

This fire ant first appeared in the United States around 1930 and steadily spread from its point of origin in Mobile, Alabama, throughout the southeastern United States. It entered Southeast Texas in the late 1960s and invaded most of the prairie chicken range by the 1970s. The red imported fire ant not only kills chicks directly, it also kills the soft-bodied insects that make up the chicks' food supply. The arrival of the fire ant within the chicken range in the mid-1970s coincides with the population decline that began at that time from a modern-day peak of 2,240 birds in 1975. This insect is a pervasive, intractable pest and represents probably the greatest modern threat to the survival of the prairie chicken. Attwater's Prairie Chicken has faced many hazards in its history; the red imported fire ant is yet one more in the life and hard times of the Texas coastal grouse.

It is difficult to know what will become of Attwater's Prairie Chicken. The perils the bird faced came long before the era of conservation and environmental concern. It lacked the public interest that was enjoyed by the bobwhite quail, wild turkey, and white-tailed deer. Val Lehmann once asked the head of the state wildlife agency why the latter did not do more to protect the bird. The answer was: politics. The public, and therefore the government, did not care, or at least did not care enough to move more vigorously on the bird's behalf. Furthermore, the prairie chicken's home, to which it was elementally bound, stood in the path of a rising society and a growing economy. Even as concern over the bird's plight grew, it took decades after Lehmann's call for action for any substantive research on prairie chicken ecology to be conducted.

The future of Attwater's Prairie Chicken lies with the captive breeding program and the continued willingness of private landowners to share their land with this endangered bird. Many things will need to go the bird's way: greater success with captive breeding, restoration of agricultural lands back to the prairie, fire ant control, more funding by the federal government, the ongoing commitment of concerned scientists and citizens to save the species from extinction, and perhaps a surcease from the hard hand of drought and stormy spring weather. The official goal of 6,000 breeding birds spread over 300,000 acres of prairie — the benchmark for removal from the Endangered Species Act — lies beyond a distant horizon. But with effort and good fortune, the bird may recover. "We have all the tools," says John Magera, "to bring about the recovery of this species." Attwater's Prairie Chicken was a vibrant part of our past; time will tell if it will be part of our future.

CHAPTER 2

Only in Texas

The World of the White-tailed Hawk

TO THE CASUAL OBSERVER travelling the long, empty roadways of South Texas — FM 1017 south from Hebbronville, say, or US 77 south from Kingsville — the bird life for which the region is renowned may not be readily apparent. It is a hot, harsh land, where prairie grass fights with thornscrub for dominance and where drought may hover over the countryside for years. For mile after mile the landscape offers struggling pastures, row crops and orchard groves, grazing cattle, acres of dense honey mesquite and granjeno, burlap grasslands scattered with solitary trees, and isolated gatherings of live oaks, like so many green-black islands in a distant sea. The land is flat to slightly rolling and parts of it, such as the long reach through Kenedy County, are studded with inland dunes of white sand, sculpted by the everlasting wind and, at first glance, oddly removed from their natural home to the east, on Padre Island and the Laguna Madre.

But beyond the fences and gates of the great ranches, among the tangles of retama and huisache along the Rio Grande that have escaped the plow, along the power lines, throughout the savannahs and oak woodlands live the birds that for decades have drawn nature watchers from throughout the US and beyond. South Texas, and especially the Lower Rio Grande Valley, is famous for its "specialty" birds — the

Clay-colored Robin, Masked Duck, Plain Chachalaca, Green Jay, Great Kiskadee, Hook-billed Kite, and dozens of other species whose ranges in the US are often largely restricted to South Texas.

Much of this bird life persists despite more than a century of land-altering encroachments of farms and ranches, cities and towns, damming and irrigation, and, in more recent times, the fences and floodlights of border control. One species, however, survives primarily because of human intervention: the White-tailed Hawk (*Geranoaetus albicaudatus*), an infrequently studied, seldom seen, and strikingly beautiful hawk whose US range lies only in the coastal counties of Texas. The hawk was previously classified under the genus *Buteo*, which represents a group of hefty-bodied, high-soaring hawks with rounded wings and short, banded tails; familiar Texas raptors such as the Red-tailed Hawk and Swainson's Hawk fall within this group. The buteos stand in contrast to the genus *Accipiter*, which includes smaller, low-flying hawks that inhabit woodlands. Cooper's Hawk and the Sharp-shinned Hawk, other common Texas birds of prey, belong to the accipiters.

Most species of America's birds of prey spend at least part of their life somewhere in Texas. Most of them inhabit the state throughout the year, while others migrate with the seasons, with peak populations in Texas occurring between September and April. A few range from West Texas to the West Coast, a few others from East Texas to the East Coast. Some raptors occur widely throughout the state, while some have smaller ranges. Others are very rare, with little pockets of occurrence, for example, in West Texas or far South Texas. Many of them have vast ranges for their annual cycle, spending summers in northern Canada and winters throughout the United States. But all of them, with the exception of the White-tailed Hawk, occur elsewhere in the US.

This bird of prey reflects the contradictory fates for wildlife that sometime accompany the intersection of modern society and the natural world. Just as rice agriculture harmed Attwater's Prairie Chicken while benefiting migrating waterfowl and shorebirds, the institution that eliminated much of the original South Texas grasslands and their avian communities — large-scale cattle ranches — today in many areas provides a stable habitat for this hawk. "There is a reason why Texas has a self-sustaining population of the White-tailed Hawk," says Dr. William P. Kuvlesky Jr., a wildlife scientist with Texas A&M University–Kingsville who has conducted research on the raptor. "If anyone deserves credit for the preservation of the White-tailed Hawk, it is the big ranches of South Texas."

Of the fifteen species of hawks that occur regularly in the US, the White-tailed Hawk has been long sought after as a Texas specialty. In its full adult plumage, this uncommon, non-migratory raptor of the Texas coastal grasslands is among our most beautiful birds of prey, with its steel-gray back, rusty red shoulders, bright white breast, and short, white tail with its inch-wide sub-terminal black band (i.e., not quite reaching the tip of the tail); the boldly marked tail is unmatched in any other US raptor. Both sexes have the same plumage and both are about two feet long (the female is an inch longer), with a four-foot wingspan for the male and a five-foot span for the female. H. C. Oberholser, in *The Bird Life of Texas*, says this hawk is "exceedingly showy — probably the most beautiful American buzzard hawk."

The White-tailed Hawk — occasionally referred to in the early literature as the "prairie hawk" — is also one of the least studied of our native raptors. Although relatively few studies on this species have been conducted, researchers at Texas A&M University in Kingsville and College Station and at Texas Tech University have revealed fundamental aspects of the bird's habitat, alarm calls, behavior, and nesting ecology in Texas. It is listed as "threatened" by the Texas Parks and Wildlife Department.

The raptors enjoy a special allure, with their aura of power and menace. A Northern Harrier banking over a plowed field; the elegant Swainson's Hawk (easily a rival to the White-tailed Hawk in aerial grace) passing through on its way to South America; Broad-winged Hawks soaring by the thousands near Galveston Bay in autumn; a Kestrel diving toward a cotton rat skittering around a winter-faded pasture: all are sights that even the most veteran Texas birdwatcher seldom tires of seeing. But the White-tailed Hawk seems to bring a special excitement.

The reason for this exhilaration eludes a straightforward explanation. Perhaps it is the bird's relative scarcity, its regal bearing atop a tall tree as it surveys the countryside, the clean contrasts of its plumage, its beauty and power on the wing. Although seeing this bird of prey in the wild is a memorable event, by the criteria of modern birdwatching it is certainly not a rarity. It does not merit a rare-bird alert, which is elicited by, say, the Aplomado Falcon or Roadside Hawk. But its appeal persists.

Once, while I was on an excursion across Aransas Bay to see Whooping Cranes, a White-tailed Hawk swept closely over the boat, its white chest and tail flashing brilliantly in the wintery sunshine. Someone shouted "White-tailed Hawk!" The great cranes we had come to see were suddenly — albeit briefly — second in interest.

"The white-tail is 'hit-or-miss' here," says Dan Walker, biologist at the Chaparral Wildlife Management Area, near Cotulla. "We are always excited to see one." Upon seeing a White-tailed Hawk during a visit to South Texas in 1900, Florence Merriam Bailey — a renowned naturalist of the time and the first woman elected as a Fellow of the American Ornithologists' Union — wrote, perhaps a bit fervidly, of "a King of Hawks looking up with calm enquiring gaze, both the gaze and pose bespeaking the silent power of the race." Seeing the White-tailed Hawk for her was "among the rare pleasures of the journey."

The White-tailed Hawk is a bird of the open country. Throughout its hemispheric range it inhabits the savannah, which are grasslands that are widely scattered with trees. From the *terrenos abiertos* (open lands) of Rio Negro Province in central Argentina, the bird ranges north to the grassy plains in Venezuela known as *llanos*, and to Peru, southeastern Brazil, and Colombia. This raptor also occurs in the pine savannahs of Central America, the coastal plains in Mexico of Sonora, Tamaulipas, and the Yucatan Peninsula, and over to the cactus-and-acacia deserts of the Caribbean. The coastal plain and barrier islands of Texas are the northernmost extent of its range.

In Texas, this tropical hawk's known breeding range extends throughout the coastal prairie from the southern counties to the Attwater Prairie Chicken National Wildlife Refuge, in Colorado County. Its range in Texas formerly extended much farther north; a breeding pair was observed in Taylor County, 150 miles directly west of Fort Worth, in 1896. Extending about 5,300 miles, the hawk's distribution from Texas to Argentina is the longest north–south range of any of the buteonine hawks.

The White-tailed Hawk was first identified in Texas by ornithologist George Burritt Sennett during a visit to the Lower Rio Grande Valley in 1878. Sennett was an example of the gentleman-naturalist so prevalent at the time, a person of independent means who devoted much of his life to exploring the natural world and making important contributions to scientific knowledge, often without formal academic training. (Charles Darwin is perhaps the supreme embodiment of this nineteenth-century individual.)

Sennett was a wealthy industrialist who manufactured oil-field equipment in Ohio and Pennsylvania. His true passion, though, was American bird life. After numerous collecting expeditions throughout the Midwest, he enthusiastically turned his efforts to Texas. On the eve of his first departure south he wrote in his notebook: "Was today studying Audubon and the Mammals and Birds of Texas and have hope of getting great quantities of valuable skins." He made three trips to South Texas and published extensively on the region's bird life. The Texas hawk was later given the name *Buteo albicaudatus sennetti* by renowned ornithologist Joel Asaph Allen, of the American Museum of Natural History; the early literature often refers to the hawk in Texas as "Sennett's white-tail."

The White-tailed Hawk's intimate dependence on the savannah and similar habitats leads it to avoid cultivated agricultural lands and densely forested or mountainous regions such as the rainforests of the Territorial Amazonia, the interior highlands of Mexico, or the dense thornscrub and oak woodlands of coastal Texas. The hawk's bond to a singular type of rangeland stands in contrast to that of some other raptors, many of which hunt and nest within a variety of habitats. The Red-tailed Hawk, for example, will inhabit mountains, forests, river bottoms, and deserts, as well as the open plains and prairies.

Savannahs and similar areas lie scattered along the Texas Coast, fragmented by the sprawl of cities and the scalped grid of farmland.

Because of this discontinuity, the White-tailed Hawk tends to have localized populations. In South Texas, the hawk's habitat occurs predominantly as the mesquite savannah, a rangeland scattered with honey mesquite and other brush species; this rangeland is especially abundant throughout the Coastal Sand Plain, a two-and-a-half-million acre region of wind-blown sand between Corpus Christi and the Rio Grande. The trees serve as nesting sites and the grass and forbs — weeds and wildflowers, such as croton and western ragweed — provide habitat for the small mammals, reptiles, and insects upon which the hawk feeds. With this prairie bounty these vast, intact, undeveloped ranches — the King, Kenedy, Armstrong, and others — provide large parcels of habitat for the hawk, yielding its largest and most stable population.

"The large ranches are a stronghold for the White-tailed Hawk," says Dr. C. Craig Farquhar, a wildlife scientist with the Texas Parks and Wildlife Department who has written the definitive monograph on the hawk. The advice that Lt. Robert E. Lee, according to King Ranch folklore, gave Captain Richard King — "always buy land, and never sell" — benefits Texas wildlife to this day.

The importance of South Texas ranches to the White-tailed Hawk rests on the changing land use that has taken place in the ranching industry over the past several decades. Ranching has long been, as renowned western novelist Elmer Kelton has written, "financially hazardous." Numerous challenges confront ranchers in their efforts to run a profitable operation: unstable livestock prices, drought, water supply, government policies, changes in consumer tastes, and costs of feed, pasturage, ranch hands, and herd health management.

Few conditions pose a greater threat to successful ranching than drought. "Drought and water supply are the major issues facing the Texas cattleman," says Laramie Adams, Director of Public Affairs of the Texas and Southwestern Cattle Raisers Association. These day-to-day difficulties are compounded by the younger generation leaving the business altogether. "They want to do something else," says Adams.

To overcome these economic roadblocks, ranchers have pursued additional ways of making money from the land; on many South Texas ranches, cattle operations no longer dominate as they once did. Oil and gas leases, wind energy turbines, ostrich and zebra ranching, agricultural crops such as sorghum, hunting leases, and the efficiencies of modern technology (computerized herd data management and DNA analysis, to name but two) have altered the face of modern ranching. Kineños and Kenedeños — the *vaqueros* who work the cattle on the King and Kenedy ranches — still ride the fences and round up cattle, but today it is more often in pick-up trucks and helicopters than on quarter horses. The evocative scene of lariat-throwing cowboys on horseback rounding up thundering herds of Santa Gertrudis, eating grub with the *patrón* around the camp fire and sleeping under the stars — so memorably portrayed by the revered Tom Lea in *The King Ranch* — is largely a vision of the past.

Hunting leases on some ranches have become a major source of income, as South Texas abounds in popular game species: wild turkey, white-tailed deer, javelina, quail, dove, and waterfowl. Leases for bobwhite quail are especially in great demand; hunters come from Europe and Latin America to pursue this popular bird, formally known as the Northern Bobwhite (*Colinus virginianus*). This grassland bird is one of the most intensively studied of our game species; Valgene Lehmann, of Attwater's Prairie Chicken fame, conducted much of the original research on the Texas bobwhite, beginning in the 1930s.

In Texas, where some of the nation's most extensive quail habitat remains, quail hunting has been popular for many years, although it has become, in the words of one hunting lodge owner, "a wealthy man's sport"; rates for guided hunts can exceed a thousand dollars per person per day at some lodges. The economic incentive to the landowner is compelling: a well-managed quail lease can generate more than three times the profit per acre than cattle ranching alone. Significantly, maintaining quail habitat benefits the White-tailed Hawk, because South Texas quail, like the hawk, need the mesquite savannah. "The White-tailed Hawk is a side beneficiary of quail management," says Dr. Kuvlesky.

The four species of Texas quail — Northern Bobwhite, Scaled, Gambel's, and Montezuma — have been declining in numbers for about forty years. Various causes have been proposed, but the loss and fragmentation of habitat most likely have led to this disquieting, relentless decline. In other parts of the country, as in the oak savannahs of Tennessee and the long-leaf pine savannahs of the Southeast, quail have suffered a similar loss. The lilting, silvery call of "bob WHITE" that at one time graced even the din of the big city has fallen silent throughout much of its historic range. South Texas ranches, with their immense, intact tracts of prairie and savannah, serve a greater need than just attending to the leisure of the affluent hunter.

Bobwhite quail require large areas of grasses and forbs for food and cover for nesting and brooding, patches of bare ground for foraging, and scattered trees and shrubs for shelter from predators, the summer heat, and the winter wind. For this six-inch-tall, six-ounce bird, the grass must not be too dense. Early- to mid-successional stages of grass are necessary to allow movement; grass for nesting must be tall enough to provide concealment from skunks and raccoons, the most common nest predators; and tall trees, such as mesquite, need smaller, low-lying brush, such as granjeno, at their base to provide sufficient cover. A pasture devoid of brush and shrubbery is not good quail habitat. A tapestry of different plant communities providing cover, seeds, and insects — tall bunchgrass here, shorter grass and open areas there, wildflowers throughout, clusters of trees and shrubs spaced at about the distance a person can throw a softball, as range managers like to recommend — best sustains the Northern Bobwhite.

South Texas ranches maintain more rangeland for quail than for any other game animal. Keeping the land suitable for quail, however, requires a vigilant management program. South Texas rangeland, like all other coastal prairie habitat, is under the constant threat of turning into a grass-choked thicket and eventually being overrun by brush and trees. To prevent this situation, land managers rely on managed grazing, brush control, and prescribed fire. These methods reflect, however distantly, the natural cycle of growth and renewal on the historic prairie.

Two hundred years ago the Wild Horse Desert, the grassland between the Nueces River and the Rio Grande, was governed, as it had been for millennia, by a natural regime of wild grazing mammals and the vagaries of weather. White-tailed deer, wild horses, and stray cattle (bison were not believed to have migrated much below the Nueces) fed throughout the prairie, roaming widely in search of fresh grass. The free-ranging herbivores would graze an area and then move on to adjacent rangeland, leaving some grass behind. Even when the Spaniards confined sheep and cattle there was little grazing pressure on the prairie. When lightning struck, or perhaps when embers escaped from a campfire, the grass could sustain a fire that would spread widely; early settlers would write of prairie fires that extended for miles. The fire would burn out the thick, dead litter and woody brush, restoring nutrients to the ground and allowing a fresh cycle of young, nutritious grass, weeds, and wildflowers while removing any invading brush. Thus the prairie maintained itself and the home for the animals that depended upon it. But the rise of the cattle kingdom altered that natural balance.

The large-scale ranching that began in the 1850s led to overgrazing of the prairie, as huge herds of cattle grazed without restraint. The prairie grass was severely damaged on several fronts: Forbs declined in abundance, the complex profile of plant species became simplified, insects, rats, mice, lizards, and snakes could not maintain healthy populations, and fuel for prairie fires diminished. Prairie fire consequently became less frequent, and the loss of fire led to the spread of mesquite, huisache and other thornscrub species. The problem of overabundance of brush became worse as cattle and birds consumed seeds and spread them around.

Cattle grazing intensified in the 1870s with the introduction of barbed wire, by which large numbers of cattle could be confined to a pasture. Later, as water supply and transportation to markets improved with windmills and railroads, cattle ranching became a huge economic enterprise that exacted a costly toll on the prairie. The pristine land of the early 1800s eventually disappeared. By the 1890s the spread of brush had become the bane of the South Texas rancher. It remains so to this day, to ranchers throughout the American Southwest. "I don't like trees," an Oklahoma rancher once sternly told this author. "Grass doesn't grow under trees."

Almost any highway through the coastal plain bears witness to the consequences of uncontrolled brush. Along Highway 239 in Goliad County, for example, live oak woodlands surround one of the last surviving bluestem coastal prairies. (This remnant Goliad prairie has been used for the private property release program for Attwater's Prairie Chicken.) Decades ago, these forests were prairie. Oak forests have their own beauty and importance, but not for creatures of the grasslands. The Texas City Prairie Preserve, a 2,300-acre sanctuary of native bluestem near Galveston Bay that has served as a refuge for introduced prairie chickens, lies next to a dense woodland of Chinese tallow that was once part of the same prairie.

The most effective tool the rancher possesses for achieving a balance between livestock and wildlife is managed grazing. As Ronnie Howard, former manager of the San Tomas hunting camp on the King Ranch's Encino Division, says, "If I have a hundred dollars to spend on quail management, I'll spend eighty of it on cattle grazing." Cattle grazing programs call for rotating cattle around pastures on a schedule and giving each pasture a year-long rest every four or five years. They have been widely used for years for wildlife management in, for example, Oregon to preserve foraging areas for elk; grazing cattle and bison on the Attwater Prairie Chicken National Wildlife Refuge serve to maintain the mixed bunchgrasses required by the endangered grouse. The Caesar Kleberg Wildlife Research Institute of Texas A&M University–Kingsville has developed much of the guidance for maintaining wildlife habitat on South Texas ranches through cattle management.

Cattle grazing programs often need to be modified to account for the erratic weather in the state's southern counties, since wet years can be followed by years of punishing dryness. Without adequate rainfall, the grasses and forbs diminish along with the insects, small mammals, and other prey that quail and hawks need to survive and reproduce. Further, it is not just the total annual rain, but also its timing, that in part determines the fate of animal populations. Sufficient rainfall in the spring and early fall bestows the greatest benefit to grassland birds, with nesting cover, forbs, and insects flourishing over the following months; quail, for example, enjoy robust populations after a well-timed rainy season. A delayed spring rain, conversely, can suppress valuable rangeland plants, even though annual rainfall may be statistically normal. During a drought, ranchers are advised to reduce stocking rates, which is the number of grazing animals per unit of pasture; overgrazing during this vulnerable time damages pastures to the point that it takes years for them to recover.

The other important method for the rancher in conserving wildlife habitat is prescribed burning — the deliberate setting of a controlled grass fire by trained professionals. Fire has not always been recognized as a tool for rangeland management. For many years, landowners suppressed fire, thinking that its destructive powers conferred no benefit to the landscape. This long-lasting attitude toward fire exacerbated the brush-spreading effects of overgrazing.

Burning must be carefully timed with suitable weather conditions of relative humidity, air temperature, recent rainfall, and wind speed and direction; prescribed fire is usually planned for every three to five years. As the fire clears out old dead grass, litter, and woody brush, the stage is set for new growth of grass, weeds, and wildflowers.

A prairie fire, when viewed up close, has a singular, frightful beauty. A prescribed burn has the look of a minor military operation. First, the fire crew huddles and reviews the fire plan — developed months in advance — and evaluates the wind conditions. Fire trucks take up positions on the perimeter, in case the fire gets out of control; law enforcement vehicles stand by to keep onlookers from getting too close. After evaluating wind direction, the fire crew selects the areas to burn first and then, outfitted with flame-resistant clothing, helmets, and radios, fan out to different locations in the pasture. They ignite their drip torches, each the size and shape of a fire extinguisher, and with sweeping motions set the dense grass ablaze. Soon a wall of lapping orange fury is chewing across the pasture, hissing, popping, roaring, kicking back a blast of heat in the face, as when one opens a hot oven door, sending up a heavy, churning curtain of smoke — some of it white, some of it dark gray or black — two, three, four hundred feet in the air. Specks of ash swirl about the observer's face and the fireplace smell of burning wood lends an odd domestic touch to the scene. The air above the flames takes on a liquid shimmer, rippling thickly like clear water flowing over a pebbly mountain stream. The smoke leans gently with the wind as it rises, and then vanishes. Long, snaky dervishes of dust and smoke — fire tornadoes — spiral high up, throwing themselves apart as quickly as they form. The fire advances and then all is quiet, an aftermath of smoking black stubble lying where the pasture once stood. One has to be reminded that this violence is an act of renewal, of restoration, without which the life of the prairie hesitates.

The great columns of smoke can be seen for miles and, soon, as if on cue, the birds of prey arrive. They come to partake of the spoils, the fleeing prey; among them are Red-tailed Hawks, Swainson's Hawks, Harris' Hawks, vultures, caracaras, and, often in the greatest numbers of all, the White-tailed Hawk.

Prairie fires famously attract *G. albicaudatus*; in South Texas the locals call it the "fire hawk." Veteran birdwatchers say that the feeding behavior around a fire will often be the most reliable way to observe this elusive raptor. Both adults and the darker, less boldly plumaged sub-adults appear, gliding in and out of the smoke, disappearing from view where the rising, wavering smoke is thick and then re-appearing before disappearing again, sailing high above the flames and then diving sharply, jousting in mid-air, flying close to each other and then darting away, with harshly flapping wings trying to chase off the perceived intruder, constantly watching the charred ground below for the escaping snake, grasshopper, or cotton rat. The smoke of a prairie fire is visible up to around ten miles away; the relatively large numbers — often two to four dozen — of this sparsely distributed hawk flocking to the fire suggest they come from considerable distances.

As the use of grazing and fire has evolved over the years, the approach to dealing with brush has changed as well. Elimination of brush — keeping any and all brush out of the pasture — was at one time the dominant approach, especially when the only desired land use was cattle ranching. Some brush, however, is necessary to South Texas wildlife. Contrary to the thinking of some early researchers, mesquite did not "invade" South Texas from Mexico when cattle ranching altered the prairie. Mesquite and other thorny, woody species have long been a natural part of the landscape; indeed, they give the countryside much of its harsh and unique allure. By pursuing a brush management strategy of selective use of physical removal and herbicides, as well as managed grazing and fire, ranch managers preserve the vital features of the mesquite savannah. As the quail flourishes on the great ranches, so does the White-tailed Hawk.

Ranch land stewardship, however, often goes beyond merely serving the needs of the hunter. "Many ranchers are true conservationists who take great pride in maintaining healthy wildlife populations on their properties," says Dr. Kuvlesky. "They do not have just an economic interest in wildlife."

The White-tailed Hawk's sharply defined habitat requirements reveal its intimate bond with the coastal savannah. Nest sites are usually in trees and shrubs that are no more than ten to twelve feet high, have small canopies, and are spaced widely throughout expanses of short-to-mid-sized grass. The pasture should have a variety of species of grass and forbs that supports the bird's broad diet of rats and mice, lizards, frogs, snakes, small birds, and large insects like grasshoppers. The hawk avoids croplands, urban areas, and overgrazed pastures and areas where tree cover exceeds around 40 percent. It builds its shallow, elliptical nest at or near the top of the tree or shrub, usually six to twelve feet above the ground. Honey mesquite, granjeno, lime prickly ash, yucca, and others in the southern counties, and, farther north on the coast, shrubs such as yaupon and Macartney rose (the primary nesting site on the Attwater Prairie Chicken National Wildlife Refuge) provide these low-lying nesting sites.

At the onset of the breeding season, in late December or January, White-tailed Hawks form pair bonds and engage in courtship with aerial displays. The pair selects a nest site and, over the next four to five weeks, builds the two-foot long nest with tree branches and grass. The female lays eggs, usually two to three, sometime between March and May and incubates them for about four weeks. The parents feed and protect the nestlings for about another seven weeks before the young hawks leave the nest.

The struggle for survival begins from the earliest moments of a nestling's life. Of the pair of young that are usually raised, one of them is occasionally smaller and less vigorous. The other, stronger nestling dominates and takes a greater share of the food that is brought to the nest by the parents. This competition was recorded by an observer who was studying a pair of white-tailed young in a nest in a yucca on the Laguna Atascosa National Wildlife Refuge, near the Laguna Madre. One of the young was visibly larger and more active than the other. The other one was so weak it was barely able to hold its head erect. When an adult brought food — ground squirrel, blue crab, ribbon snake, cottontail, horned lizard — the other nestling took most of it. Eventually the smaller one caught up to the other in size and was able to grab its share. As they both increased in size and strength, they would exercise by standing on the edge of the nest, extending their wings into the wind and briefly hovering above the nest before falling back down, in preparation for the fateful day when they would leave the nest for life on their own.

Like many hawk species, the White-tailed Hawk is monogamous, forming pair bonds for life. At the beginning of each breeding season, the pair re-establishes the bond in, as Dr. Farquhar says, a sort of "renewal of vows." The short, scattered trees common to the savannah serve not only for nesting but also for a vantage point for scanning the surrounding grassland for intruders and for prey. White-tailed Hawks are territorial — roughly two to three square miles for one breeding pair, according to the available data — and will send out alarm calls when potential predators encroach on the pair's territory. As nest predation is a major cause of mortality and nest failure, the sounding of alarm calls is an important survival strategy.

Dr. James C. Merrill, a contemporary of George B. Sennett and one of the early observers of Texas bird life, described the hawk's alarm call in his 1878 monograph on birds in the vicinity of Fort Brown, Texas (today's Brownsville). In his report on the "white-tailed buzzard," he wrote of seeing two hawks' nests, each on top of a yucca in the Palo Alto prairie, near the fort. "The nests were not more than eight feet from the ground, and were good sized platforms of twigs," Merrill wrote. "While examining these nests, the parents sailed in circles overhead, constantly uttering a cry much like the bleating of a goat."

As an opportunistic feeder, the hawk consumes a range of prey species as each one becomes more vulnerable than others; feeding at prairie fires portrays opportunistic feeding in the extreme, as escaping prey provide a ready target for soaring hawks.

Clarence Cottam, an early Texas conservationist and long-time director of the Welder Wildlife Foundation and Refuge, near Sinton, once felt the need to offer a cautious defense of the White-tailed Hawk's diet. In a 1939 survey of birds and their preferred food items, Cottam, somewhat obliquely, pointed out that "[i]t is probable that Sennett's Hawk is normally beneficial in its food tendencies." Such apologetics were necessary in an era when the attitude of many farmers and ranchers to birds such as crows and hawks was one of suspicion and hostility, because of presumed damage to crops and livestock.

The bird's prowess in spotting prey in the prairie grass speaks to the fabled power of a hawk's eye. A researcher was observing a White-tailed Hawk preening on a fence post on the Attwater Prairie Chicken refuge when suddenly the bird shot away with powerful wing beats. It flew just above the ground for about 500 feet — half again the length of a football field — before diving into the grass. The hawk then promptly returned to the fence post carrying a slender glass lizard.

Estimating the size of the White-tailed Hawk's Texas population has long presented a challenge to wildlife scientists. Part of the reason for that is, ironically, their ranching redoubt. Because so many of the hawks inhabit the large ranches, to which access is severely restricted, producing a reliable estimate of the hawk's population has not been possible. Field reports from the early 1900s, however, suggest the hawk was fairly common. The early literature refers to it as a "common resident," "not uncommon," and "plentifully distributed." A 1910 South Texas survey stated the White-tailed Hawk was second only to Harris' Hawk in numbers.

From around the 1930s, however, it appeared that the population began to decline. As habitat for nesting and foraging disappeared under brush encroachment and ever-expanding cities and farms, the bird's population, as best as it could be determined, began to diminish year after year. Pesticides are another potential culprit, as eggshell thinning studies have shown the hawks' eggs became substantially thinner between the 1940s and 1970. Although pesticides were never specifically identified in relation to the hawk, heavy use of DDT and dieldrin has been well documented from that era as a cause of reproductive crash in birds. The Brown Pelican, once almost extirpated from the Texas and Louisiana coasts, is perhaps the best-known example of this disastrous consequence.

Several decades of data from the venerable Audubon Christmas Bird Count offers some insight into the shifting fortunes of the White-tailed Hawk. Christmas counts have taken place annually for over a century across the country. Between December 15 and January 5 each winter, groups of experienced birders spend a day in the field estimating the number of individuals of all bird species that they observe between sunrise and sunset.

Although Christmas count results may not have the statistical rigor of an aerial waterfowl survey conducted by, say, the U.S. Fish and

Wildlife Service, they are important in identifying population trends from year to year. From around 1950 to 1970, Christmas count records of White-tailed Hawk sightings in South Texas reveal a steady, severe decline. Then, throughout the 1970s, the count data indicate an increase in population. A 1977 estimate placed the population at around 200 breeding pairs; the 2017 Christmas Count reported 417 hawks throughout the coastal prairie.

While the US breeding range of the White-tailed Hawk is restricted to Texas coastal grasslands and barrier islands, it does occasionally wander across the Sabine River. The Louisiana Bird Records Committee considers the bird a vagrant in Louisiana, where twenty-seven individuals have been reliably reported since 1888, mostly from the southwestern parishes of Cameron, Calcasieu, and Jefferson Davis.

The New Mexico Department of Game & Fish classifies the hawk as an "accidental," a bird far removed from its natural range and not ever expected to be present. Although reputable observers have reported the hawk several times, there have been no confirmed sightings in the state. (A rare, unexpected species is confirmed only by photography, audio recording, or by collecting a specimen.)

The breeding range once extended into Arizona, as documented with the collection of an egg from a nest and two adults, the last specimen being collected in 1899; according to the Arizona Game and Fish Department, no specimens have been obtained since then. Arizona saw the expansion of the cattle industry in the late 1800s, and the consequences for wildlife were much the same as they were in Texas. After 1900, both the Masked Bobwhite (a subspecies of the Northern Bobwhite) and the White-tailed Hawk were extirpated from Arizona as breeding species due to the exhaustion of savannah grasslands by overgrazing, followed by reduction of fire and the spread of brush. The last credible report of a sighting of the White-Tailed Hawk in Arizona was in 1965. Sightings of the hawk have also been reported from the state of Sonora, in northwestern Mexico, although there is scant documentation for its distribution or breeding status.

As throughout Latin America, the populations of the White-tailed Hawk in Texas tend to be localized and discontinuous along the coastal prairie. Although the hawk does not migrate, it appears to undergo local, seasonal movements, possibly because of climate-related changes in the availability of prey. Personnel at both the Anahuac National Wildlife Refuge, in Chambers County, and the Brazoria National Wildlife Refuge, in Brazoria County, have reported that sightings of the hawk are much more likely in winter than in summer. Prescribed burns at these and other coastal refuges routinely attract several dozen White-tailed Hawks.

The White-tailed Hawk's dependence on the tree-scattered prairie makes its future vulnerable to forces both natural and man-made. Drought drastically threatens populations, as the lack of rainfall diminishes grass cover, starving the small animals on which the hawk survives. Reduced food supply leads to lowered reproduction; studies have shown diminished nesting success — fewer young produced — in periods of drought. Although drought affects all wildlife, the population of birds that are abundant can better withstand such stress; for species with smaller populations and precise needs for food and nesting, such as the White-tailed Hawk, long-term drought may inflict greater damage.

Should quail hunting ever decline in popularity in Texas, or if the vast, storied ranches of South Texas change their land use away from maintaining wildlife habitat, the hawk could suffer accordingly if savannah disappears. Moreover, there is the ever-present specter of expansion of cities, farms, and highways throughout the Texas Coast that will forever remove trees and grass from the natural scheme of things. If the open country survives in coastal Texas, however, so should the White-tailed Hawk.

LINDA M. FELTNER

CHAPTER 3

South to Aransas

A Crane's Odyssey

April–September, Wood Buffalo National Park, Northwest Territories, Canada

IT HAD BEEN A good summer, finally, for the young Whooping Crane family of the Sass River. For the three years that they had been mates, Buff, the male, and Regina, the female, had nested each spring in the wetlands that embrace this narrow tributary of the Little Buffalo River, in the Northwest Territories portion of Wood Buffalo National Park. But each year, they had failed to raise a chick.

Their first attempt at nesting, when Buff and Regina were both five years old, resulted in two infertile eggs. The following year two chicks hatched, the usual two days apart. But while they were still confined to the nest a hailstorm struck the area and both chicks perished. A year later two chicks again hatched, but the older chick pecked his nesting mate to death in the act of siblicide that is common in the Whooping Crane.

It had been very dry in the park that third summer. With water levels low throughout the wetland ponds, the crane family had to wander farther afield to find food, having to get close to one of the dense green, tree-lined ridges that meander among the ponds. As the parents were foraging one afternoon with the surviving juvenile, a lynx darted out from the tamarack and attacked the youngster. Buff and Regina both lunged at the predator, pecking at him frantically, trumpeting their alarm calls. But it was too late; the young crane was dead. Buff and Regina quickly flew away to safety.

But this summer had been different. Once again, Buff and Regina had returned together to the park in mid-April from their winter home on the Aransas National Wildlife Refuge, on the central coast of Texas. They had mated near Meadow Lake, in Saskatchewan, toward the end of spring migration back to the park. Soon after their arrival Buff selected a nest site in a pond near their previous nesting locations, several miles from the nearest neighboring crane, on the north bank of the Sass River. Their parents had fledged more than a decade earlier in this same region.

Now, as a breeding pair, they had returned each year to raise their own young within the roughly two hundred square miles of forest and ponds lying between the Sass River and, twelve miles to the north, the Klewi River. Their companion cranes were nesting throughout the northeast corner of Wood Buffalo National Park, including north of the park boundary, but they mostly used the Sass–Klewi wetlands, where the nesting ponds (generally only an acre or two and rimmed with bulrush, cat-tail, and sedge) are crowded edge-to-edge by the thousands, like water lilies on an East Texas lake. Their pale green waters spring from the nearby Caribou Mountains, filling the shallow basins through the cracked bedrock. From this rare landscape comes the world's only self-sustaining wild flock of migrating Whooping Cranes.

Buff and Regina soon set about building their nest — half the size of a dining-room table — in the center of a bulrush island set within a shallow, acre-sized pond. Nesting conditions were good: Water levels in the ponds were not too deep from heavy snowmelt and not diminished from drought. Summer fires, common in the boreal forest, had spared their nesting area.

Regina laid two eggs and the month-long incubation passed without incident, with each parent spending hours at a time, all day and night, atop the nest. As the time of hatching approached, Regina would spend more time on it than Buff. He would stand watchfully nearby, taking over the nest several times a day to allow Regina to feed. The four-inch-long beige eggs hatched in early June, two days apart, and two wet but soon-to-be-fluffy chicks tumbled out, each the color of a dusty pumpkin.

The first chick, Sundown, was the larger of the two and ate more robustly than did his sibling; the second chick was smaller and weaker, lacking the two extra days of growth that Sundown had been given. Late one afternoon, a cold front — not uncommon in the erratic summer weather of sub-arctic Canada — blew across the region. The following morning, many of the crane nests throughout the region suddenly held only one live chick, rather than two. The four-day-old Sundown survived, but the second chick did not.

Sundown grew quickly as his parents brought him food, and by one week of age he was strong enough to leave the nest and walk and swim. He followed his foraging parents closely, awkwardly at first but with increasing strength and height each passing day. Buff and Regina watched over him as they walked through the bogs and marshes, slowly lifting their long black legs to move about as they sought frogs, small fish, berries, and insect larvae, all the while keeping a wary eye on the forested ridges that meandered among the ponds for a lurking bear or wolf. In moments of danger, Regina would lift a wing and Sundown would huddle under it. As the danger passed, he would emerge and walk about again. Sundown's parents fed him throughout the day, giving him water beetles and fat dragonfly larvae; they would feed him for almost a year, throughout fall migration and the winter in Texas, up to the time he would start living independently the following spring.

The three cranes wandered every day along the edges of ponds in search of food, leaving a visible trail of tracks in the spongy organic muck that formed the base of each pond. After the family left the nest, they never returned to it. After each day's foraging, they roosted for the night in the shallow pond where they had ended their daily efforts. Food had been abundant, and all three cranes were in good trim.

By early September, Sundown, now in the streaked, white-and-rusty-brown plumage of the juvenile, was almost as tall as were his nearly five-foot parents. Most important, he was now capable of sustained, high-altitude flight, for the time to migrate to Texas was drawing near. Buff and Regina lingered in the area to feed Sundown a while longer, availing themselves of the abundant food, but soon the autumn weather would begin to press upon them. Daylight, which by late June in north-central Canada extended almost nineteen hours a day, was daily growing shorter. The weather was quickly turning colder and the land-hardening pall of winter would soon arrive. By the end of October all the remaining cranes would be gone, finally driven out by the harsh weather.

A long journey awaited the Sass River family; Wood Buffalo National Park is as far from the Texas Coast as London is from Cairo. And although migration is the briefest phase of a Whooping Crane's annual cycle, serious hazards lay in the birds' path. Unreliable food supplies, the vagaries of the weather, unfamiliar stopover terrain, collisions with power lines, and ignorant, careless hunters all bedevil the migrating Whooping Crane; not every one of the tall, stately white birds nesting throughout the park would survive. The cranes would also need the forbearance of property owners, since many of the roosting and feeding sites they would use on the migration corridor were privately owned.

One mid-morning in the first week of October, a strong north wind arose; a light snow had fallen the day before but the morning sky was bright and clear. The sun was warming the landscape and thermals began to rise. The crane family braced itself and stood still for several minutes, gazing up at the sky. Then, following their instincts inherited from millennia of ancestors, they lifted off on their seven-foot wings, flap-flying briefly before beginning a slow rising spiral, first Buff, then Regina, then Sundown, all circling slowly as they ascended to around a thousand feet. Then, one by one, they peeled off and, in close V-formation, headed south. Their great journey across Alberta, Saskatchewan, Montana, the Dakotas, Nebraska, Kansas, Oklahoma, and Texas had begun.

Early October, Northeastern Alberta–Southwestern Saskatchewan

The three cranes, aided by good tail winds and clear skies, flew along the path of the Athabasca River, but the next day deteriorating weather — swirling rain, snow, and poor visibility from a low-pressure system — forced them down around the Birch Mountains, about fifty miles south of the park. These conditions persisted, and the cranes were forced to stay in the area for four days, feeding and roosting throughout the small lakes and ponds. This was not a good start to their travels; typically, the flight to Saskatchewan, their next destination, takes only one or two days. Clear skies and favorable winds eventually returned and the cranes were on their way once again, arriving two days later in the southwestern corner of Saskatchewan, the neighboring province.

Buff, Regina, and Sundown spent ten days near Reward, Luseland, and Neville, a line of small agricultural towns set within a landscape of grassy, gently rolling hills, river valleys, and groves of aspen scattered with red barns, white grain elevators, and thousands of ponds and lakes carved by long-vanished glaciers. They wandered about the grain fields each day, feeding on the remains of that summer's wheat and barley harvest and roosting in nearby shallow ponds at night. The cranes would spend more time in Saskatchewan than on any other stopover in fall migration, as they built up nutritional reserves for the long journey south.

Sundown was slowly learning to feed on his own, as he pecked about the stubbly field for grain; being less efficient at foraging, he would take longer to feed than his parents, who would stand by watchfully for predators or encroaching humans. The family foraged and roosted mostly alone, gradually making their way south. Although other small groups of cranes, especially sub-adults — the one-to-four-year-old birds that have not yet bonded with mates — would sometimes join them for a few days. Flocks of their only North American relative, the Sandhill Crane, were frequently feeding nearby. As all migrating Whooping Cranes would do throughout the remainder of their migration, Buff and Regina selected ponds and farmlands for feeding and roosting that had long, open views, far from buildings or tree lines for better vigilance against predators and intruders.

They gradually made their way south through the farmlands. Around the third week of the month, the crane family set out on their final leg of migration, a rapid flight through the Great Plains to Texas. Now they would embark upon a routine, as weather allowed, of long-distance flights during the day and, by sundown, feeding and roosting, since cranes seldom fly after dark.

Late October–Early November, Montana–Texas

Buff, Regina, and Sundown departed Saskatchewan and reached the Poplar River, in northeastern Montana, the next day. The following day a strong tail wind and clear skies allowed them a 10-hour, 460-mile non-stop flight southeast over the Dakotas — often at altitudes exceeding 5,000 feet — before arriving near the Valentine National Wildlife Refuge, in northern Nebraska. As weather permitted, the cranes rode the thermals upward in a great spiral and then glided on fixed wings downwind for a half-dozen miles or so, gradually losing altitude down to around 1,000 feet before catching the next thermal updraft and spiralling upward again. They continued the next day, but rain and fog forced an early stopover near Broken Bow, in the central part of the state. The poor weather lifted the following day and the cranes then turned directly south; good weather accompanied them across Kansas. A small group of sub-adults joined them in flight over Kansas for a few hours but the younger birds eventually veered off on their own. Poor visibility once again caused a delay as they reached northern Oklahoma, and the crane family had to wander about the Cimarron River valley, making very little progress. It took another day to reach Byers Lake, in Texas, northeast of Wichita Falls and just south of the Red River. Strong headwinds came up and the crane family was again delayed, this time for four days.

When north winds returned, the cranes took to the air. After departing Byers Lake, they crossed Texas quickly. They stopped overnight near Rosebud, south of Waco, and the next day near Tivoli, northeast of Rockport. The following morning they headed southwest, and at long last the dark, tangled oak forest of Blackjack Peninsula came into view. Later that morning, a month after their departure from Wood Buffalo National Park, Buff, Regina, and Sundown were feeding on blue crabs and wolfberries in the bays and marshes of the Aransas National Wildlife Refuge.

On the last day of 1937, President Franklin D. Roosevelt issued Executive Order 7784, designating 47,000 acres of Blackjack Peninsula, a marsh-fringed wedge of land between San Antonio and Aransas Bays, on the Coastal Bend of Texas, as the Aransas Migratory Waterfowl Refuge. Aransas — which at the time had eighteen wintering Whooping Cranes (*Grus americana*) — was the second federal wildlife refuge in Texas. (The first was Muleshoe National Wildlife Refuge, which was established in 1935, sixty miles northwest of Lubbock.) A later Presidential Proclamation designated 13,000 acres of open waters surrounding the peninsula as closed to migratory bird hunting to protect the cranes. A Civilian Conservation Corps camp was soon set up at Aransas to build roads, trails, and freshwater ponds; a lone stump of a flagpole overlooking San Antonio Bay today is the only relic of that difficult time.

The creation of the Aransas refuge — renamed the Aransas National Wildlife Refuge in 1940 — was not an isolated act of environmental stewardship undertaken by the administration of Franklin D. Roosevelt. The dozens of these national wildlife sanctuaries that were established during his time in the White House — twenty-seven in 1935 alone — bear witness to an aspect of Roosevelt's leadership that has not received the wider recognition that it deserves.

In the early years of his administration, and despite the national crises of economic collapse and the gathering clouds of global war, Roosevelt had the foresight to enact programs and legislation that benefit the environment to the present day. New farm policies created by the Agricultural Adjustment Act gave financial incentives to farmers to preserve natural landscapes rather than cultivate them, a precedent for the valuable conservation easements of today. The Soil Conservation Service was established. And the Migratory Bird Hunting and Conservation Stamp Act was enacted, generating hundreds of millions of dollars over the years for preservation of waterfowl habitat through its federal duck stamp program. Much of the work of two landmark programs of the New Deal, the Civilian Conservation Corps and the Works Progress Administration, was directed toward restoring and enhancing parks and forests.

The president's action on Aransas came near the time when the Whooping Crane stood at the precipice of being removed from the roster of the world's living animals. A year after the refuge's creation, the North American population of the wild crane — in effect, the world's population, since it occurs nowhere else—was estimated at twenty-nine birds in the first census ever taken for the species. The cranes were divided at that time between a non-migratory breeding flock in southwestern Louisiana, near White Lake, and a migratory flock that spent the summer in Canada and the winter on the coasts of Texas and Louisiana.

Most of the Texas birds resided along the eastern edge of Blackjack Peninsula, with its dense, green-and-tan carpet of saltgrass, saltwort, and sea ox-eye set within an intricate archipelago of small islands, salt flats, inlets, sloughs, ponds, bays, and tidal channels. A few cranes inhabited neighboring areas, such as the adjacent barrier islands of San Jose and Matagorda. A hurricane struck Louisiana in 1940, destroying over half the cranes there; the Aransas flock declined to fifteen birds the following year, its lowest number ever for the refuge.

From a legal perspective, the Aransas refuge was not created specifically for protection of the Whooping Crane. The Executive Order simply states that the new federal facility will serve as "a refuge and breeding ground for migratory birds and other wildlife"; the order does not mention any specific species. But wildlife scientists had long been aware of the primacy of Blackjack Peninsula and its surrounding waters for the wintering crane. "It is no accident that the area selected for the Aransas National Wildlife Refuge," wrote Robert Porter Allen in his famous monograph on the crane, "was the major wintering ground for more than 60% of the Whooping Cranes then alive . . . . The refuge was set up just in time."

What Valgene Lehmann was to Attwater's Prairie Chicken, Robert Porter Allen was to the Whooping Crane. The Pennsylvania-born Allen spent three decades as a field biologist with the National Audubon Society, studying hawks, herons, and the Roseate Spoonbill, among others. But his major contribution to American ornithology lay with his work on the Whooping Crane. He spent years studying them at Aransas and following them up their migration route (what was then known about it) and into Saskatchewan. He soon recognized the difficulty of studying this wary, challenging bird. Of his first contact with the cranes at Aransas he wrote: "I remember that those first two birds seemed very far away . . . distant — but not only in a physical sense. Their arrogant bearing, the trim of their sails, as it were, would intimidate the most brash investigator. I reached our cabin that first night feeling very humble and not too happy. And that . . . is a very good way to begin." His 1952 monograph *The Whooping Crane* is a rare masterpiece of spare, elegant prose conveying dense, sweeping detail; it laid the foundation for all subsequent research on the crane. The Whooping Cranes alive today owe their existence in no small measure to Robert Porter Allen.

Although the Whooping Crane was never abundant — Allen estimated the mid-nineteenth-century population at 1,300 to 1,400 birds — it did occur widely throughout the continent, from the Atlantic seaboard and Gulf Coast to the Great Salt Lake and from the Arctic coast to the high plateaus of central Mexico, almost 1,000 miles south of El Paso. Fossil evidence from California, Florida, and Idaho suggests an even wider distribution across North America during the Ice Age.

From their nesting grounds in Saskatchewan, Alberta, and Manitoba and the Upper Midwest — Iowa, Illinois, Minnesota, and North Dakota — cranes migrated in winter to Mexico and to the coasts of Texas, Louisiana, Georgia, South Carolina, and New Jersey. They formerly ranged from Utah, Wyoming, and Wisconsin to Arkansas, Missouri, and Alabama. The prairie marshes of northern Iowa were especially dense in nesting habitat. The early naturalists, such as Mark Catesby (who provided the first record of the crane, in 1722), Alexander Wilson, and John James Audubon, wrote accounts of these great white birds. Lewis and Clark, on their famous journey, provided the first record of the crane in what is today North Dakota, the heartland of its historic range.

In Texas, Whooping Cranes were scattered along the grasslands and marshes of the coast, from the Sabine River to the Rio Grande delta. Their occasional occurrence inland led to the first Texas record, in 1845, when Col. George A. McCall of the US Army reported a Whooping Crane pair at Arroyo Hondo, near San Antonio.

Southwest Louisiana, however, probably had the most concentrated population of Whooping Cranes on the Gulf Coast. The state's tallgrass prairie with its ponds and sloughs and its coastal region with its lakes and marshes provided an abundance of habitat that supported both a migratory, wintering population and a resident breeding flock. The migratory birds used the inland prairies and cheniers — ridges that are relics of ancient beaches — while the resident flock inhabited the marshes near White Lake and Grand Lake, southwest of Abbeville. Early settlers and trappers told of seeing cranes throughout the region over the decades following the Civil War. But after the war the inevitable clash between an expanding American society and its wildlife would play itself out with disastrous consequences for the Whooping Crane.

Beginning around 1870, wildlife scientists and conservationists watched with dismay the ever-shrinking world of *G. americana*. Year after year, decade after decade, the disturbing facts had increasingly pointed to a decline of a species on a continental scale, as the bird disappeared throughout most of its historic range. The draining and plowing of the prairies, the coming of the railroads, the rise of the Gulf Coast rice industry, and the constant pressure of human settlement, agriculture, and hunting all contributed to the perilous decline.

Prior to the Migratory Bird Treaty Act of 1918, which protected migratory birds, hundreds of cranes were shot, mostly during migration; Nebraska, whose Platte River valley is a major migration stopover today, was the scene of many killings. The toll that hunting took on the bird was particularly grievous since, in addition to its small population, the crane does not reproduce quickly; adults do not breed until the age of three, but more commonly at age four or five. Crane pairs usually produce only one chick each year. Chicks mature slowly and not all of them survive the storms and predators of the nesting grounds or the hazards of migration. Most likely, the documented shootings make up a small fraction of the actual killings. In 1918, a Louisiana rice farmer by the name of Alcie Daigle — his name deserves to be remembered — shot and killed twelve Whooping Cranes on the inland prairie; they were feeding on waste grain near his thresher. Whooping Cranes would eventually disappear altogether from Louisiana.

By the early 1920s, their known nesting grounds were largely wiped out in both Canada and the United States. The last nest in the US was reported near Clear Lake, Iowa, in 1894; the last known Canadian nest, in Saskatchewan, was reported in 1922. As nesting grounds were destroyed, the wintering cranes likewise declined in number. The winter months formerly saw cranes from New Jersey to Central Mexico, a distance of some 2,000 miles. But by the mid-twentieth century, a twelve-mile stretch of marsh and bay on the central Texas Coast was the sole surviving remnant of the cranes' vast former range. There was a general belief that the Whooping Crane was standing at the abyss of extinction.

As Robert Porter Allen pointed out, the establishment of the Aransas refuge brought the public's attention to the desperate plight of the crane and a sharp focus on the need for action to save the bird. The cranes of the refuge posed a formidable question, however: Where did they go in the summer? Whoopers would arrive at Aransas in October, stay for the winter, and then depart in April. They would return to Aransas the following October, sometimes with one or two juveniles. Clearly, they were nesting somewhere, presumably in the vastness of Canada. But where exactly? No one knew. Allen put it bluntly in his monograph: "The location of the present breeding area is unknown."

For wildlife agencies and conservationists to develop a plan for long-term survival, researchers needed a fuller understanding of the cranes' needs for food and habitat. The establishment of the refuge at Aransas was the crucial beginning of the effort to save the magnificent bird. There the crane could be accurately counted and protected from human intrusion under the watchful eye of the U.S. Fish and Wildlife Service; its daily activity and behavior could be monitored and its needs for winter food and territory assessed; its reproductive success for the year, measured by the number of juveniles arriving in the fall, could be measured. If fewer cranes returned in the fall than departed the previous spring, researchers needed to know why.

At the end of the Second World War, wildlife biologists embarked upon the Whooping Crane Research Project, an international effort between the US and Canada to locate the birds' nesting area. Elemental questions about the nesting area first had to be answered. Where did the cranes nest? What kind of landscape did they need? Was the nesting area near human settlement and perhaps in danger of intrusion and disturbance to the famously wary cranes? Was it in the midst of an agricultural region, in the path of a tractor that would soon enlarge an adjacent wheat field? Or — it was desperately hoped — was the nesting area in a remote, inaccessible area with abundant habitat and free of human threat? The Whooping Crane Research Project set out to answer these questions.

For the next decade, officials with the U.S. Fish and Wildlife Service and the National Audubon Society, with assistance from the Canadian Wildlife Service, carried out an almost Arthurian quest over the cranes' historic Canadian nesting region and beyond. Each year they covered tens of thousands of miles by small aircraft, carefully watching the land below for the big white birds. They traversed the Canadian prairies, across the Northwest Territories, British Columbia, and the Yukon Territory, all the way up to Point Barrow, at the far northern edge of Alaska. Nesting cranes were never seen; occasionally a report of a crane sighting from a local farmer or hunter was received, but nothing would come of it. The searchers kept going.

Finally, in the summer of 1954, a pair of cranes with a juvenile was spotted from a Canadian forestry service helicopter investigating a fire in the northern part of Wood Buffalo National Park near the Sass River. While this was the best lead the searchers had in nine years, they needed to confirm the nest site on the ground. Plans were immediately set in motion to send out a search party the following summer to set up a base camp a mile or so from where the cranes had been seen from the air (a journey well over a hundred river miles) and then hike up close to them.

The ground search team, led by Robert Porter Allen, set out by canoe up the Sass River. They quickly found out such an approach was impossible. The narrow, twisting Sass — in the southern US the river would be called a "creek" — confronted the team with logjams and beaver dams, in addition to the constant torment of flies and mosquitoes. They turned back after a week.

They hiked back to Fort Smith, the nearest town, and quickly arranged for a helicopter to take them to the nesting area. Allen and his search partners were struggling though the dense spruce and birch of one of the ridges when they saw a flash of light from the pond surface beyond and, walking in the shallow water, two adult cranes. A few days later the searchers saw two juveniles with them. The long, arduous search was over. It was precisely the kind of natural realm everyone had hoped for. "It is in truth a lost and unknown place," Allen observed, "and the nesting whoopers should continue to prosper here in the spring and summer months, as they have evidently done for so many years."

The location of the breeding grounds of the Whooping Crane had become a mystery that captivated the public, in both the US and Canada. This awareness had arisen from a massive publicity campaign in both countries — a component of the Whooping Crane Research Project — as researchers had enlisted the aid of the public in observing cranes both in the nesting region and along their migration path. The plight of this great bird had become a matter of national concern; a syndicated Sunday morning cartoon even once featured the struggling crane. Civic groups, Boy and Girl Scout troops, outdoor and nature clubs, sportsmen's organizations, farm clubs, garden clubs, state wildlife agencies, and the broadcast and print media were all enlisted to report sightings of cranes. "[E]very man, woman and child in Saskatchewan," Allen wryly noted, had heard of the Whooping Crane.

On September 20, 1954, an article entitled "Here Come the Cranes" appeared in a national publication. The author, West Texas native John O'Reilly, was writing about this great wildlife detective story. "The exact location of the nesting grounds of the remaining whoopers has not been found," the author wrote. "This summer a scientist hovering in a helicopter over the wild country of Canada's Great Slave Lake looked down and spotted four Whooping Cranes and a young one. His find was the best evidence so far of the general location of the breeding grounds." The publication was not the *National Geographic*; it was not *Audubon Magazine*; it was *Sports Illustrated*. The following summer that same publication proudly announced: "North America's greatest ornithological puzzle has been solved." The *Saturday Evening Post* published articles about the bird and the *New York Herald Tribune* dispatched a correspondent to meet the search party on its return. With the establishment of the winter refuge at Aransas, the discovery of the nesting grounds in Wood Buffalo National Park, and wide public concern over the fate of the bird, the stage was now set for the crane to step back from the precipice of extinction.

November–March, Aransas National Wildlife Refuge

By late November, the last of the migrating cranes were on the ground at Aransas. Not all of the birds departing Canada, however, survived migration. One adult was shot by a hunter in Nebraska; another adult collided with a power line in Kansas, on a low-altitude take-off on a foggy morning. And one three-year-old crane — a sub-adult, which had been at Aransas the year before — did not return and was unaccounted for.

The Sass River family, some eight decades after the founding of the refuge, was now part of a gathering of more than four hundred cranes. After a brief stopover near Mustang Lake, on the northeast corner of Blackjack Peninsula, Buff, Regina, and Sundown flew over to Matagorda Island. Here they would spend the winter, in a territory bordering Espiritu Santo Bay that Buff had claimed years earlier after he had pair-bonded with Regina; it was close to the area where he had spent his first winter with his parents. In previous years Buff and Regina defended this same haven of around 500 acres from intruding, neighboring cranes. Now, with a chick in tow, the intensely territorial wintering cranes defended their territory — essentially, their winter food supply — with even greater vigor.

The last cranes to arrive that year were a female and a juvenile; a male — possibly the adult that died on the power line in Kansas — did not accompany them. By the time they reached Aransas cranes had already claimed most of the prime feeding grounds along the costal marshes. They attempted to work into a winter territory near Sundown Bay, but resident cranes chased them off each time. Eventually, the single parent and fatherless young crane had to retreat to the uplands of Blackjack Peninsula until they could find some marsh that wasn't being defended.

Each day Buff, Regina, and Sundown walked throughout the marsh, slowly lifting their long black legs, stepping carefully, jabbing at the mud, shallow water, and marsh grasses for crabs, clams, worms, and snails with their large, powerful beaks, their rich red crowns bright in the sunshine and the unruly shock of tail feathers — known as the "bustle" — busy in the wind. They pecked at the wolfberry bushes, grabbing at the clusters of the red, marble-sized fruit that had grown plump during the fall. By the end of January, most of the wolfberries would be gone.

The crane family wandered about their territory throughout the day, searching for food as the wind, tides, and rainfall constantly re-sculpted the face of the shallow waters and grass that made up this marginal world between land and sea. In December and January, the seasonal low tides and north winds would drain tidal flats, forcing blue crabs into deeper water elsewhere. The cranes would then move over to shallow bays and channels and feed on clams and snails. When the flats became refreshed later with rain or tides, the crabs — and the cranes — returned.

Ponds or tidal channels that became too deep — two feet or so — kept cranes away. On rare occasions when freezing weather hardened the shallow waters of the marsh, the cranes moved into the bluestem and oak uplands of Blackjack Peninsula to feed on acorns, grasshoppers, and crayfish. Whenever refuge personnel would conduct a prescribed burn in the uplands, cranes would invade to forage, frequently in the presence of White-tailed Hawks soaring overhead.

Sundown stayed closest to Regina while foraging; just as he had learned migration from his parents, they were teaching him to feed himself. Regina gave him food, but he also pursued crabs and poked at wolfberries more frequently on his own. Buff stood somewhat apart from them, feeding while watching over the territory with a stern vigilance. Sundown seldom ventured off alone, as Buff and Regina constantly tended to him.

The crane family spent each night in the middle of a broad, shallow pond — allowing a better view of an approaching bobcat, coyote, or alligator — with their heads tucked under a wing, becoming ghostly, contorted apparitions in the gathering twilight. At dawn the next day they stretched out in the morning sun and began again their constant search for food.

Neighboring cranes would occasionally approach the "boundary" of the Sass River family's territory. Buff, observing the encroachment from a distance, would sound a loud challenging call — *ker-lee-oo!* — and then extend his neck and take off on a low, quick flight toward the intruder. Both cranes would jab their beaks at each other, flapping their wings and jumping up and down, slashing away with their feet, then staring at each other, standing still a few feet apart on either side of the skirmish line. The face-off would end after a few minutes. The intruding crane would eventually back down and leisurely stride back into his own territory.

The weeks and months passed. The cold gray days of January and February slowly gave way to the gentler touch of the approaching spring. Sundown grew and matured under Buff and Regina's constant care. He was almost entirely white by now, with only a light wash of juvenal cinnamon brown on his head. Sundown now faced the next great challenge in the life of a young crane, as the time to return to Canada was drawing near.

The Whooping Crane avoided, ever so closely, the finality of extinction, but its journey from the precipice was slow and halting following the establishment of the refuge. As documented by the annual aerial census that has been conducted at Aransas since 1950, the flock would increase by a few birds for a year or so and would then decline. Nesting failure, predation, hunting, the hazards of migration, and unknown causes resulted every several years in fewer birds returning to Aransas in the fall than departed in the spring. Then the following few winter seasons would see slight increases. This ebb and flow persisted for decades (as it does today), with the population hovering around twenty to thirty birds. The 1950s — the "time it never rained," as Elmer Kelton called Texas's historic drought — was especially hard on the Whooping Crane, as the population showed very little growth.

But despite the all-too-frequent loss of cranes, the wild migratory flock did slowly increase, reaching forty birds by the mid-1960s and sixty birds by the mid-1970s. The Aransas–Wood Buffalo flock took a full half century since the creation of the Aransas refuge to reach 100 birds, all descendants of the historic low of fifteen cranes that were present there in 1941. In all but two years since the founding of Aransas, juvenile cranes — sometimes only two or three, sometimes more than thirty — arrived each fall with their parents. This slow but steady recovery would not have happened without the legal protection that the crane has long enjoyed: the Migratory Bird Treaty Act of 1918 and the 1966 Endangered Species Protection Act, the precursor to the 1973 Endangered Species Act, which protects *G. americana* today.

The 1980s ushered in an era of hope for the crane, as its population doubled, to almost 150 birds, in nine years. "It was a decade of great optimism," says Tom Stehn, a leading crane expert who served for fourteen years as the Whooping Crane coordinator at Aransas and who was conducting the crane census at the time.

In earlier times when crane numbers were low and good habitat was plentiful, the cranes were able to defend territories as large as 800 acres in size. But as more cranes returned to Blackjack Peninsula each year, they could not defend areas as large as they had previously, because neighboring cranes on their territories prevented them. As a result, crane pairs and families crowded into increasingly smaller areas, reaching an apparent lower limit of about 250 acres per territory as the marshes of the original refuge became fully occupied. Then, as they outgrew their mid-century cradle along the peninsula's eastern edge, they moved on to adjacent areas to find habitat of suitable acreage rather than continue to push into even smaller areas. Meanwhile, sub-adult cranes find foraging areas as they can, where they will not be chased off by defending adult cranes. Sub-adult cranes often range near the first winter territory that they spent with their parents.

Cranes began to establish territories first on Matagorda Island, then on San Jose Island, and, later, on the Lamar Peninsula (west of Blackjack Peninsula across St. Charles Bay) and Welder Flats (north of the peninsula across San Antonio Bay). As they established new territories next to old ones, their winter range expanded within these areas. On rare occasions cranes spend time on inland landscapes where they can find shallow, open water; in recent years Whooping Cranes have wintered at Granger Lake, northeast of Austin and in Wharton County, southwest of Houston. But coastal habitats make the best crane wintering territory.

Today, Whooping Cranes occupy the bayside marshes of Matagorda north almost to Port O'Connor; a few years ago a crane family was observed south near Port Aransas, to the great excitement of the community. In contrast to the twelve miles of the crane's 1930s range on Blackjack Peninsula, the wintering crane now occupies more than fifty miles of the Texas Coast. Today more cranes inhabit Matagorda and

San Jose islands than were found on the original refuge. The Aransas National Wildlife Refuge itself has grown to almost 115,000 acres and now includes Matagorda Island and a portion of the Lamar Peninsula.

However, we should not necessarily call the saga of the Whooping Crane a "success story," for the crane's fight for survival is far from over. The decades of conservation efforts have certainly brought the crane to its healthiest numbers of modern times; the winter survey of 2017–2018 reported 504 cranes at the Aransas National Wildlife Refuge. But any animal species with only around 850 or so living individuals (the Aransas–Wood Buffalo flock in addition to cranes in introduced flocks and in research facilities) is still a species very much at risk, especially a species that reproduces as slowly as does the Whooping Crane. There is, moreover, a pragmatic implication to how the story of the crane is viewed. "I become uneasy when the Whooping Crane is called a 'success story,'" Tom Stehn says. "'Success stories' tend to receive less funding."

The Whooping Crane now confronts a new challenge: Where will future generations of this demanding, intensely territorial bird live in winter, as its numbers continue to rise? Wildlife scientists recognize that the availability of suitable wintering habitat on the Texas Coast will determine in large part the future of the wild migratory flock. Nesting habitat is abundant in Canada — Wood Buffalo National Park is larger than Switzerland — so a shortage of nesting landscape is not anticipated to limit the flock's growth, although mortality of sub-adult young in Canada is now recognized as one of the leading causes of loss of cranes. The *International Recovery Plan for the Whooping Crane* has established a long-term target of 1,000 cranes for the Aransas flock, in the event that introduced flocks elsewhere in the country (also part of the recovery plan) fail to survive. But will there be enough additional coastal habitat in Texas for the migratory-flock crane to meet this goal, which is expected to take at least twenty-five years, if not much longer?

The recovery plan, developed by the Canadian Wildlife Service and the U.S. Fish and Wildlife Service, set this 1,000-crane number as the population that would give the bird not only better stability to withstand a serious assault on its winter habitat but also greater genetic diversity. This population target also represents a "down-listing" goal, which would allow the crane to be moved from the "endangered" category to "threatened." A population level that would allow a complete removal of the Whooping Crane from the threatened or endangered status has not been determined.

An array of potential disasters hovers over the Whooping Crane: prolonged drought strangling the San Antonio and Guadalupe rivers, which feed Blackjack Peninsula's embracing bays; a Category 2 hurricane scouring Matagorda Island (as Hurricane Ike did to Bolivar Peninsula in 2008; similarly, in 2017 Hurricane Harvey struck San Jose Island, south of the refuge, flooding the marshes but not damaging them); a benzene spill on the Gulf Intracoastal Waterway poisoning the marshes. (Ninety percent of waterway cargo are petroleum products and other industrial chemicals.) The crane is especially vulnerable to such events today because of its confined wintering range; a larger population of cranes spread over a longer reach of the coast will be better able to recover from a destructive event. But a population of a thousand cranes will need to seek winter grounds thirty, forty, or even fifty miles beyond the environs of Aransas.

Tom Stehn, one of the principal authors of the recovery plan, estimates that the coastal bays and marshes roughly fifty miles north and south of Aransas, as well as Aransas itself and its immediately surrounding areas, currently have sufficient suitable habitat that could ultimately support around 1,100 cranes. Cranes would need to move into areas such as the Nueces River delta, the southern reaches of San Jose Island, the mainland west of Matagorda Peninsula, and locales near the mouth of the Colorado River to reach the 1,000-crane target.

But the cranes' need for new locations for winter foraging places the bird squarely in the path of the projected human population growth — an estimated 30 percent for the Coastal Bend alone — that is anticipated for coastal Texas over the next forty years. Condos, marinas, housing, commercial development, roadways, resort hotels, wind farms, power lines — not to mention the huge demands on river flow, the lifeblood of the estuary — will add to the loss of almost 75,000 acres of natural wetlands that the Texas Coast has already endured over the past forty years.

The increases in sea level and temperature predicted by climate change models would, if the forecasts hold true, render many coastal areas unusable by the crane. Some projections call for a sea level rise of up to four feet along the coast by the end of the century. If this happens the marshes would become too deep for the crane, making food items such as crabs and clams unavailable to the bird. During geological time, the coastline gradually moved inward when sea levels rose as the Ice Age glaciers melted, maintaining the marshy environment. But unlike today, there was no human infrastructure to impede the migrating shore line. Coastal development in many areas today will not allow room for the marshlands to move farther inland.

Furthermore, with the slow increase in temperature there are fewer hard freezes along the coast, which have historically kept mangroves south of Port Aransas, well away from crane areas. Mangroves do not provide good crane habitat, but they are steadily spreading north and now occur as far north as Matagorda and Galveston bays. Between 1990 and 2010 the spread of this tropical, salt-tolerant tree increased by 4,000 acres, primarily along the upper coast and often at the expense of salt marshes. Wolfberries, another important staple of crane winter food, would decline in abundance with rising marsh salinities brought on by higher temperatures.

Cranes need marshes and shallow bays not only that provide food but also that have long unobstructed views and that are, ideally at least, a half-mile or so from human activity. This openness was an important endowment of Aransas: not only its relative remoteness from human activity but also the lack of tall trees nearby that could conceal predators. In fact, no plants taller than a crane occur near the refuge marshes. Most coastal marshes on the eastern US coast, in contrast, have much taller vegetation. The Guadalupe River delta, just northeast of Aransas, is probably less suitable for crane use because of its stands of tall reeds.

Coastal areas that will adequately serve the crane will most likely become increasingly scarce. The Whooping Crane is now entering a new era of dependence upon the private landowner, since less than ten percent of suitable unoccupied habitat along the Texas Coast today is public land. (Public lands, however, do play a vital role in the cranes' life, as shown by their heavy use of migratory stopovers such as the Quivira National Wildlife Refuge, in Kansas, and the Salt Plains National Wildlife Refuge, in Oklahoma.) Moreover, as Wade Harrell, the Whooping Crane Coordinator at the Aransas refuge, points out,

"As cranes move away from the refuge, we can't manage them as well." Fortunately, many landowners — the large landowners, at least — welcome the great bird.

"The Whooping Crane does not have the stigma of most endangered species," Harrell says, referring to the cranes' increasing occurrence on private lands, not only on the coast but also on upland properties. "The large landowners near Aransas like to see Whooping Cranes on their property. People have learned to live with the crane. And the crane is more of a generalist in its habitat than we once thought."

As with the White-tailed Hawk in South Texas, large landowners have played an important role in the flocks' stability. John Welder of Victoria, a fifth-generation Texan whose farm and ranch family endowed the Rob and Bessie Welder Wildlife Foundation, has had cranes for twenty-five years on the 4,000 acres of marshes on Welder Flats, on the east side of San Antonio Bay across from the refuge. "My personal experience with cranes is very good," he says. Welder strongly advocates conservation efforts such as the Wetlands Reserve easements program of the National Resource Conservation Service, by which landowners of marshes and other wetlands are compensated for restoring and maintaining the natural landscape for the benefit of wildlife. "Good stewardship of the land is important," Welder says. "We can't just develop everything along the coast."

Stehn writes that, because recovery is expected to take decades at least, reaching a population of 1,000 Whooping Cranes remains "a distant goal." Whether or not habitat distant from Aransas will be usable decades from now or even if cranes will seek it out remains a huge unknown.

For all its sleek elegance, the Whooping Crane is not a delicate creature. It is a tough, adaptable bird that has overcome much hardship. "The species is very resilient," says Wade Harrell. "They have been through some difficult times, such as the drought of the 1950s." And while the threat of predation constantly hovers around them, Whooping Cranes don't go down without a fight. Captain Tommy Moore, of the Fulton-based tour boat *Skimmer*, always likes to say, "When I see a crane face off with a coyote, my money's on the crane."

But setbacks still fall upon the wintering crane community, as the massive die-off in the winter of 2008–2009 sadly demonstrated. That winter, twenty-three cranes — almost 9 percent of the flock — died at Aransas. The carcasses that were recovered showed emaciation and evidence of predation, indicating possible food stress. As birds become weak from inadequate nutrition, they are more susceptible to disease from intestinal parasites or viruses and are less able to resist attacks by, for example, bobcats and alligators.

Such peaks in mortality on the refuge are not rare, as similar events in 1990–1991 and 2005–2006 reveal. While it is difficult to determine the exact cause of death in these die-offs, crane researchers have identified a consistent pattern: The years when crane mortality is unusually high have been marked by lower than normal rainfall the previous summer and fall. Reduced rainfall leads to less fresh water in the rivers which then creates higher salinities in the bays; this affects the food supply of not only the cranes but of all animals that depend on the brackish world of the estuary. Blue crabs are the crane's most important winter food; cranes can consume as many as eighty blue crabs a day, according to Dr. Felipe Chavez-Ramirez, formerly of Lake Jackson's Gulf Coast Bird Observatory and the world's leading authority on the diet of the Whooping Crane. While the crustacean can tolerate saline extremes, it becomes abundant only in intermediate salinity.

Crab numbers have been declining on the coast for thirty years, due to many factors, according to coastal fisheries experts: diminished freshwater inflow from rivers, enhanced predation by a well-managed finfish population, poor water quality from run-off from the mainland, loss of grassy habitat from coastal development, and abandoned crab traps (which number in the tens of thousands). Since the late 1980s, commercial crab harvesting has declined more than eighty percent. Despite all of this, the resourceful Whooping Crane has still managed to increase in number.

But crabs do more than just provide the bulk of the cranes' nutrition; they lie at the heart of what makes Aransas so important to the life of this bird. Cranes, which can consume a variety of food items, could possibly survive without crabs or with a greatly reduced crab population, but their daily struggle for life on the refuge — not to mention their return to Canada and successful nesting, which require abundant energy reserves — would be much more difficult. "Crabs allow the crane to concentrate within the marshes," says Dr. Chavez-Ramirez. "If the crane had only snails and clams to eat, it would have to spread over larger territories." If territories had to become larger, the landscape would support fewer cranes. As Tom Stehn says, "Cranes come to Aransas for two reasons, the marshes and the crabs."

Attempts to establish self-sustaining flocks elsewhere in the country with captive-raised cranes have shown a modest degree of success, despite some severe setbacks. Unfortunately, attempts to establish a migratory flock between Idaho and New Mexico and a non-migratory flock in central Florida were unsuccessful. On a more optimistic note, a population known as the Eastern Migratory Population is "very promising," according to Liz Smith, a Texas-based crane researcher with the International Crane Foundation. This flock — numbering around 100 birds by 2019 estimates — was introduced into Wisconsin and was trained to migrate to the Florida Gulf Coast with ultra-light aircraft. These cranes have nested and successfully fledged chicks, several of which have completed migration. However, growth in the flock has been very slow because for them, as for wild whoopers everywhere, predation is a constant menace.

The genetic "bottleneck" that the crane went through in 1941 presents another obstacle to its long-term survival. The entire current population of Whooping Cranes has descended from the fifteen cranes at Aransas from that time; the gene pool of the Louisiana flock has been entirely lost. Modern research reveals that only six to eight cranes of the early Aransas flock have provided the genetic material for cranes alive today. This enormous loss of diversity — the assemblage of different genes that supports animal populations in their struggle to thrive amid the hardships of nature — is especially concerning for a small population like that of the Whooping Crane.

In 2011, a non-migratory flock of ten juveniles was introduced in southwest Louisiana, near White Lake, where whoopers occurred historically in great numbers. Five years later, two chicks successfully hatched, the first wild-hatched Whooping Cranes the state has seen in seventy-five years. The program is showing continued signs of success, with five chicks hatching in 2018. Despite the constant threats of predation and hunting, crane scientists recognize that there is abundant crane habitat in Louisiana, with its vast marshes, prairies, rice lands, and crayfish farming.

The fragility of these two introduced flocks starkly reinforces the importance of the expansion of the wintering range of the Aransas–Wood Buffalo cranes. But the vulnerabilities of that native wild flock also lend urgency to the need for the introduced flocks to become self-sustaining. For the crane no longer to require the protection of the Endangered Species Act, much work, and many years, lie ahead.

Fortunately, it is easy to care about the Whooping Crane. With its great height — it is the continent's tallest bird — its brilliant white plumage, crimson crown, and enormous black-tipped wingspan, its sweeping journey over the Great Plains twice a year, its pugnacity and protectiveness, its life story filled with drama and danger, the crane has enjoyed a great advantage over many companion species in the melancholy gallery of America's endangered animals. In fact, the world's fifteen species of cranes have long held a special place in myth and cultural folklore as harbingers of longevity, harmony, and good fortune. Author Peter Matthiessen calls them "the birds of heaven."

From the earliest days of the Aransas National Wildlife Refuge — indeed, from the early 1900s when the lengthening shadow of extinction began to spread over the Whooping Crane — this species has drawn from the public a level of interest and concern seldom granted other animals. *Grus americana* has stood at the forefront of American conservation efforts for more than a century.

It is, however, reasonable to ask: Have the decades of effort and cost supplied by two nations to save the crane been worth it? For that matter, should we attempt to preserve any endangered species, plant or animal? The answer is, of course, yes, but that answer comes down many avenues of human experience. Some would argue for our moral obligation, as stewards of the earth, not to allow any species to be pushed into extinction. Some would speak of aesthetics and the need for natural beauty; others would call for saving species as essential to the ultimate preservation of the human race; still others would point to the living world for its treasure of knowledge and sources of medicine and other substances that benefit humankind. Perhaps there is also some ineffable belief, something that eludes language or capture by data and dollar signs, some collective inner voice that tells us, quietly and simply, that a world with cranes — and prairie chickens and redwoods and wolves and butterflies — is better than a world without them.

Despite the validity of these lines of reasoning, however, there is the stubborn fact that endangered species that have demonstrable economic value will ultimately engage the public most effectively for their preservation. "The best form of conservation is when you can attach economic value to the species," says Robert Love, former director of Louisiana's crane reintroduction program with the Louisiana Department of Wildlife and Fisheries. In this category the Whooping Crane has no equal; this bird is big business.

The Whooping Crane was the original model for nature tourism in the United States. Beginning in the 1970s, Whooping Crane tours out of the Rockport–Fulton area, and the attendant industries such as hotels and restaurants, have become a multi-million-dollar enterprise for the Coastal Bend. "No other single bird species has had the impact on nature tourism in the US as the Whooping Crane," says Victor Emanuel of Austin, Texas, one of the nation's preeminent nature tour leaders.

In the late 1990s Al and Diane Johnson, owners of the Crane House bed and breakfast on the Lamar Peninsula, purchased an 840-acre tract of land on the peninsula that had been used for decades by wintering cranes. The Johnsons sold a third of it, including all the crane marshes, to The Nature Conservancy, which then donated that parcel to the wildlife refuge. The remaining portion is now protected by a conservation easement from the conservancy for which the Johnsons received compensation. In return, the land is guaranteed in perpetuity to be used for nothing but wintering crane habitat. "We are all about the cranes," Mrs. Johnson says.

She tells the story of the time a female crane on her property suffered a broken leg, possibly from an alligator attack. The male then became intensely protective, staying closer to her than usual, slowly circling her all day to keep predators at bay while she fed, ensuring that she safely returned to the sheltering marshes each night. This convalescence went on for several weeks. Her leg finally healed, and the crane pair was able to depart for Canada in the spring. (On another occasion, a guest started disturbing the cranes in an attempt to get them to fly. Mrs. Johnson summarily evicted him.)

The lives of cranes and humans can sometimes intersect in unexpected ways. One day in the 1980s a small aircraft was following a flight of three Whooping Cranes during a study of the bird's fall migratory behavior. The birds were approaching Dallas from the north and it became apparent to the researchers on board that the cranes were about to enter air space at the Dallas/Fort Worth International Airport. The pilot contacted air traffic control and the airport — today the nation's fourth busiest — instructed an approaching American Airlines flight to hold steady at 22,000 feet until the cranes safely passed below.

Early April, Aransas National Wildlife Refuge–Wood Buffalo National Park

And so the winter passed for the Sass River family. Buff, Regina, and Sundown spent their days wandering about their territory on Matagorda Island, feeding around the marshes and bays, fending off the occasional intruder from their little piece of winter haven, roosting at night in the open shallow waters. Buff began to engage in the antic courtship dance with Regina — leg raising, running and jumping, wing flapping — reinforcing the cranes' lifelong pair bond. Sundown had matured toward young adulthood, learning to feed himself but still under the protective eye of his parents. Their companion cranes around the refuge went about their daily business as well, indifferent to the daily tour boats packed with binocular-toting tourists and festooned with long camera lenses lined up like artillery on a gunship. Tugboats with their chuntering growl plied the Intracoastal Waterway, slowly pushing barges that made great Vs in the water that splashed up against the edge of the marsh. One crane had been killed by an alligator; no cranes died from ignorant hunters. Overall, it had been a good winter for the Aransas–Wood Buffalo flock.

Then one bright morning in early April, all three cranes became restless and started to feed very vigorously, unlike their typical leisurely manner. This went on for several hours, eventually trailing off to a few occasional jabs in the mud. Suddenly, Buff went into an alert posture — standing tall and motionless, staring straight ahead — even though no predator or intruding crane was in sight. Regina and Sundown soon followed with this alertness, and all started preening and stretching their wings. They each took a long drink of water. Buff then tilted his head to the side, a seldom-observed behavior during routine crane life, as if to get a better look at the sky. The sun had warmed the marsh and the rippling heated air began to move skyward. Still in their full alert posture, the three cranes gathered close together with Buff in the lead. He leaned into the southeast wind, stretched out his neck, took a few running steps, and lifted off. Regina, and then Sundown, followed close behind him. The three cranes spiralled a thousand feet in the air and then veered just slightly northwest. Their journey back to Wood Buffalo National Park had begun.

Their spring journey to the nesting ground would be hurried, unlike the month-long fall migration the previous October. The long, mild days of the Canadian boreal forest summer, with their abundance of fish and insects, pass quickly. For Buff and Regina to have time successfully to nest again, they needed to return promptly to the park. However, shortly after leaving Aransas they encountered stormy weather and made it only to the small town of Thrall, just northeast of Austin. The weather improved the next day and they reached Byers, near the Red River, where they spent the night. The next day they covered over 400 miles, avoiding a stopover in Oklahoma before reaching northwest Kansas, but stormy weather kept them on the ground there for three days. They finally reached central Nebraska but the worsening weather again delayed the cranes, this time for four days. The weather finally cleared and the cranes headed north once again, spiralling and gliding, spiralling and gliding, flying two, three, sometimes four to five thousand feet in the air. A series of one-day stopovers each in South Dakota, North Dakota, and Saskatchewan brought them to a boreal pond near Fort McMurray, Alberta, not far from the southern boundary of the park.

The morning after they arrived in Alberta, the crane family fed around the pond for a short time. Then Buff and Regina suddenly stood still, braced themselves against the wind, and took off. Sundown did not join them but stood quietly and watched as his parents flew out of sight. He stayed behind and fed a while longer, but then, with slow, strong wingbeats, he took off and headed north, a lone white crane flying into the blue Canadian sky. Sundown was now on his own. Buff and Regina had brought him to young adulthood, but now the task of building a new nest lay before them. Their life as a family had come to an end.

Buff and Regina returned to the Sass River, close to where they had nested the year before. Sundown would spend the summer farther north, near the Nyarling River. The following October, Sundown, in the company of a small flock of sub-adults, and Buff and Regina, with a new chick, all returned to the Aransas National Wildlife Refuge.

CHAPTER 4

Redhead Nation

Winter on the Mother Lagoon

THE WATERFOWL OF NORTH America, in all their strength and beauty, with their ebb and flow across the continent with the passing seasons, are the glory of our native birdlife. Great squadrons of honking geese; swans, whose grace and serenity conceal a fierce combativeness; rainbows of ducks — drakes in breeding plumage with their race-car iridescence of red, blue, green, and purple, hens in their muted, monastic browns — make up the continent's forty-five regularly breeding species of the bird family Anatidae. In summer, they nest and raise young throughout the boreal forests and river deltas of the Arctic and Alaska and the wetlands across the northern United States and the Prairie Provinces of Canada — Alberta, Saskatchewan, and Manitoba. In winter they migrate, by the tens of millions and over thousands of miles, to the bays, estuaries, and marshes of the Atlantic, Pacific, and Gulf Coast, and to open water and wetlands throughout the American interior and Mexico.

Each species has its own story, its own arc of birth and growth and death, its own realm within the natural world that it stakes out for itself. But when considered in view of the lives and needs of this great avian family, one species, the Redhead (*Aythya americana*), stands apart. In summer this powerful diving duck with the distinctively rounded, coppery-red head — the Aleuts call it Ka-vē-im-much, or "round-head" — nests and raises young throughout the millions of freshwater ponds that occupy the Prairie Pothole Region, an area of the northern Great Plains half the size of Alaska. But in winter, Redheads — not all of them, to be sure, but most of them — migrate to only one place, a long, narrow, salt-laden lagoon on the northwest coast of the Gulf of Mexico, the Laguna Madre of Texas and Tamaulipas, Mexico, a coastal bay system of abundant finfish and meadows of seagrass. The vast winter ranges that most other ducks seek out, such as the Mississippi Alluvial Valley, the Mid-Atlantic Coast, and the Central Valley of California, do not serve this species. The Redhead descends by the hundreds of thousands each autumn, predominantly to the Texas laguna; in flight at a distance their massed, swirling, black-and-silver-gray bodies appear, as former *Houston Chronicle* outdoor writer Shannon Tompkins describes, like "smoke on the horizon."

Like other ducks, the Redhead feeds in summer on a variety of insect larvae and aquatic plants — musk grass, sago pondweed, and many others — offered in abundance by the prairie pothole wetlands. In winter, however, their diet narrows, for the most part, to the rhizome — an underground stem that gives rise to shoots that emerge into the water — of one species of seagrass, shoal grass (*Halodule wrightii*). While other ducks feed on many types of plant and animal life in winter and can acclimate to changes in food supply, the Redhead does not. "It is rare among waterfowl for a single food item to be so predominant in the diet," says Dr. Thomas C. Michot, a coastal ecologist with the University of Louisiana at Lafayette.

This sharply defined bond with land and food is seldom encountered in American waterfowl. "The Redhead is unique in that its winter distribution is extremely concentrated in a relatively small geographic area where it forages on such a simple diet, primarily shoal grass rhizomes," says Dr. Bart M. Ballard, a waterfowl researcher with the Caesar Kleberg Wildlife Research Institute at Texas A&M University–Kingsville. This extreme dependence confronts the Redhead, which occurs only in North America, with an uncertain future. The Laguna Madre of Texas, like other areas on the Gulf Coast, is facing an array of assaults, both natural and man-made. Shoal grass grows in large meadows only under specific conditions of water quality, and those conditions prevail, in North American coastal waters, primarily within the Laguna Madre.

The last several decades, however, have seen a degradation in water quality in the Texas laguna, largely from the Gulf Intracoastal Waterway, which runs through the bay system, and from ever-increasing sewer discharge from the cities and towns of South Texas. As we will see, the mid-twentieth-century construction of this economically important waterway, and its maintenance requirements today, have led to continuing changes in water circulation, nutrients, turbidity, and salinity that severely diminish the ability of shoal grass to thrive. This problem has been aggravated by the continuing population growth of South Texas and the accompanying increase in municipal and agricultural run-off through the Arroyo Colorado, which drains into the laguna. As a result, the once-abundant meadows of the ducks' food supply have been slowly retreating. But if shoal grass disappears, where will the Redheads go?

The Redhead's journey to the Texas Coast begins from its breeding range within the open, un-forested wetlands of the northern Great Plains, Prairie Canada, and the Great Basin. Waterfowl nest in a broad range of habitats — wetlands, grasslands, forests, tundra, offshore rocky islets — and the Redhead, like most other species, has specific requirements for this crucial step in its annual cycle. The Redhead needs lakes and ponds deep enough to support stands of emerging aquatic plants — cattails, bulrushes, sedges, and others — at the water's edge. At some point in its life history each waterfowl species requires a wetland in one of its many forms: swamps, prairie potholes, playa lakes, river floodplains, bogs, fens, vernal pools, wet meadows, ponds, bays, and coastal marshes. Although Redhead nests have been recorded over much of the continent, the greatest number of them breed within two regions: the Prairie Pothole Region, a glacier-sculpted area of the north-central US and south-central Canada, and, to a lesser extent, the marshes of the Great Salt Lake, in northwestern Utah. Most of the Redheads that winter in Texas come from these two breeding ranges.

The Prairie Pothole Region extends in an irregular arc from the American Midwest deep into the Prairie Provinces of south-central

Canada. The region produces 60 to 80 percent of North America's waterfowl game species, such as the Canvasback and Mallard, as well as the Redhead. Over half of the continent's total duck population — estimated to vary annually from twenty-five to forty-eight million ducks — nests within this region, even though the area constitutes only ten percent of the total waterfowl breeding area of North America. This 300,000-square-mile region produces more ducks than does any other place on earth.

In spring the potholes, most of which are less than an acre, fill with rain and snowmelt. Plant communities of reed grass, river grass, sedges, cattails, rushes, and other aquatic plants soon emerge along the water's edge. These wetlands, which lie throughout the pothole region's grasslands and parklands, provide ducks with nesting habitat, food, and refuge from predators. During periods of low rainfall, many ponds dry up; in rainy periods, ponds fill with standing water and become useable for breeding. Known as "May ponds," they annually range in number from three to eight million.

The Redhead arrives on its breeding grounds by April, having departed its wintering range in South Texas and elsewhere by the end of March. The return to the breeding grounds is a speedier, more urgent journey — the Redhead flies faster than most waterfowl — while the autumnal southward journey is more drawn out. There is no need to reach the wintering grounds as quickly as possible, but there is an advantage to a rapid return north. "A fast return north gives the ducks a better selection of nesting sites, more abundant food, and more time to re-nest if the first nest fails," says Dr. Ballard. The journey north for most migrating birds is more direct and therefore quicker; they are more concentrated, flying at night and at a higher altitude, stopping at less frequent intervals for feeding. In fall, the southbound migrants are spread out and tend to follow the contour of the land, taking a less direct flight path along rivers and around mountains.

Many duck species are divided into two broad categories, based on feeding behavior: diving ducks and dabbling ducks. Redheads are diving ducks, which are adapted to diving for food by having legs set farther back under their body, allowing for more powerful swimming underwater. This group includes the Canvasback, Lesser Scaup, and Ring-necked Duck. Diving ducks are also known as pochards, a strange term whose etymology remains obscure. Dabbling ducks (also called "puddle ducks"), in contrast, have more centered legs and feed by up-ending themselves, submerging only the upper body. This group includes the Northern Shoveler and the most populous and widespread North American duck, the Mallard.

Like most waterfowl, the Redhead is seasonally monogamous; a male mates with one female but leaves her as soon as she starts incubating. The following year will bring a new mate. Breeding pairs, formed by their elaborate courtship ritual on the winter range, migrate together; mating occurs either during migration or after arrival in the north. The hen then sets about the tasks of building her nest and laying eggs between early May and mid-summer. The ducklings grow, feeding on grass and insect larvae, and are ready to fly after about eight weeks. The young are secretive in their habits, lurking among the nests or on the lake margins. Redhead hens care for their chicks up to the time they are ready to fledge, or fly off on their own. This is typical of waterfowl; as described by H. Albert Hochbaum in his classic *Travels and Traditions of Waterfowl*: "[I]n ducks the parent-child relationships are short, and family traditions are thus weak."

The breeding season is a time of renewal but also a time of high danger. Nest failure — the failure of a nest to produce even one hatchling — is often extreme. For each nest that yields ducklings, seven or eight or nine may be destroyed, yielding none at all. The causes are numerous and ever-present. Predators are the most severe hazard facing North America's waterfowl: Mammals such as the red fox, striped skunk, and raccoon, and birds such as crows, hawks, and owls will all prey on eggs and nesting hens. Storms and floods wipe out nests and scatter eggs. The nesting hen may simply abandon the nest, for reasons that are not always evident. Potholes may dry out in a drought and the farmer's plow may take its toll. When land is plowed, habitat is chopped into smaller parcels, rendering nests more accessible to predators. Every young of the year that survives to make fall migration is a story of success over adversity.

The Redhead often puts off her nesting duties onto another hen. This species is notorious for parasitic nesting, the reproductive behavior by which a female lays her eggs in the nest of another brooding hen, forcing that hen to raise her chicks to the detriment of her own. Parasitic nesting is observed in many species of birds — cuckoos and cowbirds are among the best known — but this behavior is most pronounced in waterfowl, especially the Redhead. Hans Christian Andersen's famous story "The Ugly Duckling" is, among other things, a tale of nest parasitism. It tells the story in which all but one of a clutch of eggs hatch together, with all the cute little ducklings looking alike. The late-hatching egg, however, produces a chick that is "too big, and peculiar-looking." The ducklings make fun of him and chase him away; he grows into a swan.

In late summer and early fall, the adults and the young of the year depart their breeding range for the onset of fall migration. By October, they are gathering on the Laguna Madre in increasing numbers, reaching their peak by the end of the year. A U.S. Fish and Wildlife Service aerial survey in 2007 showed a population of more than one million Redheads on the Texas laguna, almost 90 percent of all Redheads surveyed along the Gulf of Mexico. For six months the Laguna Madre — the Mother Lagoon — will be their refuge from the northern winter.

Redheads will also migrate to other areas on the Gulf Coast for winter feeding. "Shoal grass occurs in other areas along the coast," says Dr. Michot. "But it isn't predominant in these areas as it is in the Laguna Madre." The sparse shoal grass meadows of Chandeleur Sound, in Louisiana, may be the winter home for up to 20,000 Redheads. The Florida Coastal Bend region, between Panama City and Cedar Key, has seen as many as 80,000 Redheads; the Mexican laguna will have even greater numbers, when shoal grass is available. The greatest number of Redheads, however, goes to Texas.

Seagrass is the Laguna Madre's endowment to the Texas Coast, and to the Redhead. These marine flowering plants occur along tropical and temperate coasts worldwide, including all three US coasts. Highly specialized for underwater growth and propagation, seagrasses rank among the most valuable of coastal habitats, comparable to salt marshes, coral reefs, and mangrove swamps in their importance to marine life. Awareness of seagrass and its significance, however, have been relatively slow to enter the arena of public concern; other marine habitats, such as coral reefs, with their colorful and exotic creatures, long ago captured the public's interest. Between 2002 and 2012 *National Geographic* magazine published 166 items on coral reefs and only two on seagrass.

Although found in most Texas bays, seagrass grows most abundantly within the Laguna Madre, where ideal conditions of temperature and elevated salinity have historically prevailed. It covers well over half of the bay bottom, in meadows of tens of thousands of acres, from Corpus Christi Bay to South Padre Island. Foot-long, narrow leaves gently sway in the shallow, pale green water and the dark green masses, when viewed from above, are almost a spectral, elemental presence. The five Texas species (shoal grass, manatee grass, turtle grass, clover grass, and widgeon grass) form communities that serve as a refuge and nursery for young fish, shrimp, and crabs, a food source for waterfowl and sea turtles, a filter for removing suspended sediment, a source of nutrients from decomposition, a stabilizer against erosion of the bay bottom, and a sheltering habitat for communities of algae and invertebrates that live upon their stems and leaves.

These aquatic plants, moreover, are the basis for the Laguna Madre's bounty of wildlife; the bay's renowned community of waterfowl and finfish — flounder, spotted seatrout, red drum, black drum — would not exist without them. The Texas Parks and Wildlife Department routinely monitors finfish; although the bay represents only about 20 percent of coastal bay surface in Texas, it produces more than 50 percent of the state's finfish catch. The laguna's seagrass, which represents about 80 percent of all the state's approximate 170,000 acres of seagrass, is a resource unlike any other along the Texas Coast.

Geography and climate converge at the Laguna Madre to create the conditions necessary for seagrass to thrive. Seagrass needs clear, warm, shallow water, with salinity levels near those of the open ocean, in areas of low rainfall and high evaporation rates that are protected from offshore wave energy. The 277-mile-long Laguna Madre of Texas and Tamaulipas consists of long, narrow basins with an average depth of around three feet, largely isolated from the open Gulf of Mexico by barrier islands. The climate of hot, dry summers provides scant input of fresh water from rainfall or storm water run-off from the mainland. The 115-mile-long Texas laguna is divided into two bays of approximately equal length, the upper laguna and lower laguna, that are separated by a land bridge, known as the Land Cut, that reaches from Padre Island to the mainland.

The mainland adjacent to the Texas Laguna Madre is largely undeveloped, with the Kenedy and King ranches and the Laguna Atascosa National Wildlife Refuge forming much of the western boundary of the lagoon. Surface run-off, so abundant from urban settings, has historically been minimal from these areas, helping to preserve water clarity; in recent years, however, municipal run-off has increasingly damaged water quality. The bay bottom is primarily siliceous sand, which tends to remain settled out. Water circulation is restricted and gulf tides do not have much influence on the Laguna Madre.

Wind, however, is a constant presence on the coast of South Texas. The persistent and strong south and southeast winds for which the region is famous results from the regional geography. High-pressure systems over the western Gulf, which circulate clockwise, combine with a semi-permanent counter-clockwise low-pressure system over the Central Mexican Plateau and the Sierra Madre Oriental. These two systems cause an airflow that moves north into the middle and Lower Rio Grande Valley, creating, in the words of the National Weather Service, the Valley's "ceaseless wind."

Wind moves water to one side or the other, generating one of the laguna's distinctive features, wind tidal flats, while distributing organic matter throughout the water column. "Astronomical tides are very restricted in the Laguna Madre," says Dr. Paul Montagna, a coastal ecologist with the Harte Research Institute at Texas A&M University–Corpus Christi. "Wind has more influence on water level and water movement there than anywhere else on the coast."

No rivers today flow directly into the Laguna Madre of Texas. The Rio Grande, which flowed into the southern end of the lower laguna about four thousand years ago, connects to the gulf south of the Brazos Santiago Channel, amid the clay dunes — lomas — and the muddy gray remoteness of Boca Chica, the southernmost reach of Texas. The Rio Grande's flow has been so choked off over the years that gulf water is known on occasion to move up the river channel with the tide, temporarily creating an inverse estuary. In fact, the mouth of the river closed in 2001. The rivers that connect to the Mexican laguna — the Rio San

Fernando and the Rio Soto la Marina — are relatively short waterways and do not carry much flow unless a storm occurs.

Most of the mainland inflow into the lower laguna comes through the Arroyo Colorado, a relict waterway of the ancient Rio Grande, and the recently man-made North Floodway; these channels are the main drainage system for the urban storm and agricultural water run-off of the Lower Rio Grande Valley region. Unless rainfall is substantial, however, freshwater outflow is scant, although it has been steadily increasing as the Valley grows in population, creating greater run-off from towns and farms.

The Laguna Madre of Texas and Tamaulipas is situated, moreover, within an anomalous pattern of precipitation; the region to the north, the upper Texas Coast, and to the south, the Yucatan Peninsula, each receives more than twice the average rainfall of the Laguna Madre's twenty-five to twenty-six inches a year. Since more water evaporates from the laguna than enters it, bay water becomes saltier than the adjacent gulf, acquiring its rare and defining characteristic of hypersalinity. The gulf has an average salinity of 35 to 36 parts per thousand (ppt); today the laguna's salinity can routinely reach 40 to 50 ppt or even higher, depending upon drought and rainfall. There are only four other known hypersaline lagoons in the world: one on the Crimean Peninsula, one in Baja California, and two in Australia; of these embayments, the Laguna Madre of Texas and Tamaulipas is the largest.

The lower laguna is less saline than its sister lagoon to the north, the upper laguna, because of the greater input of gulf water through the Port Mansfield Channel and Brazos Santiago Pass, at the far southern end. In addition, the Arroyo Colorado and the North Floodway bring in fresh water from the mainland. The upper laguna lacks such connections and exhibits generally higher salinity. This distinctive geography sets the Laguna Madre apart from the typical Texas bay systems of the upper and middle coast. Those estuaries have wider connections to the Gulf, are deeper, and are fed by rivers carrying large volumes of fresh water, sediment, and nutrients into them, diluting the bay water to salinity levels much below those of seawater.

By the end of the year, the Redheads that will spend the winter on the Laguna Madre will have all arrived after their journey of almost 2,000 miles from Utah, the Dakotas, and Prairie Canada. Over the winter months they will feed throughout sandy, shallow flats where shoal grass is abundant. As wind tides move water around the bay, deep areas become shallow and shallow areas become deep; Redhead flocks will move about as necessary to feed in shallow waters.

They feed for about half of the daylight hours, usually starting right after sunrise up until mid-day. Dipping their heads or tipping over in the shallow water, they plow their beaks into the sand and silt, ripping and eating the shoal grass's rhizomes with their sky-blue, black-tipped beaks. The shredded leaves float to the surface and wash up in long, ragged drifts along the shoreline, piled up like seaweed on a sandy beach in Galveston. Shoal grass rhizomes make up around 85 percent of the Redhead's winter diet, with widgeon grass accounting for another ten percent, and small snails and clams associated with the grass community making up the rest.

At rest, Redheads flock by the thousands, in long meandering skeins of red, black, and silver-gray, floating quietly on the water. An approaching disturbance — a recreational fishing boat, perhaps, or a patrol boat of the Texas Parks and Wildlife Department — will send them exploding into flight with rapid, powerful wing beats, wheeling and turning in tight formations before sharply descending to the water. They are swift and adroit on the wing, but upon landing with a brief tumble, their clumsiness reveals itself. Unlike some species that can shoot practically straight up from the water — the Northern Pintail, for example — the Redhead needs a running start before going aloft; this is characteristic of the diving ducks, with their set-back legs. The market hunters of the early 1900s exploited the Redhead's tendency to land on water in compact flocks; with their huge five-shot pump guns and semi-automatic shotguns, they could kill a dozen birds in seconds. Since Redheads lack the natural wariness of many ducks, this probably contributed to the ease of their being hunted.

Since the mid-twentieth-century, the Laguna Madre's seagrass has been undergoing a gradual transformation in distribution and a decline in abundance. Human activity, both within the bay and on the mainland, has led to alterations in water circulation, salinity, and clarity. These changes, although protracted over decades, have led to a change in the conditions essential for all species of seagrass growth. Dense meadows have become sparse meadows; sparse meadows have disappeared, leaving only open bay bottom. Although shoal grass once dominated the lower laguna, today it has receded to a small fraction of its former range because manatee grass and turtle grass — worthless to the Redhead — have largely replaced the shoal grass within the lower bay.

In 1949, an event occurred that would set this slow transformation in motion: the completion of the final segment of the Gulf Intracoastal Waterway, between Corpus Christi Bay and Brownsville. This 12-foot-deep, 125-foot-wide shallow draft canal sliced through the entire length of the Laguna Madre, including the Land Cut. The commercial value of the entire Texas waterway is indisputable; data from the Texas Department of Transportation show that eighty-six million short tons of cargo were carried in 2014, valued at approximately $29 billion and representing almost 60 percent of the total tonnage recorded over the entire gulf waterway system. This segment of the waterway, however, came at a cost. Its construction brought about a fundamental change in the way water circulates within the Texas laguna, a change that was to have enormous consequences for the seagrass community.

Before the waterway's construction, the upper and lower lagoons were isolated from each other because the mid-section Land Cut was intact. Communication with the gulf was tenuous: The lower lagoon had only Brazos Santiago Pass, at its southern end; the upper lagoon was indirectly connected to the gulf through Corpus Christi Bay, at its northern end. And in the hot, windy world of the Laguna Madre, bay water routinely reached the high levels of salinity — 80 ppt or greater — that suppressed seagrass growth. Ecologists believe that the pre-waterway laguna probably lacked seagrass over most of the bay, even though the first systematic grass survey did not take place until 1957.

Despite the prevailing high salinity at the time, some seagrass most likely did occur in isolated, localized areas: shoal grass and widgeon grass in both lagoons and manatee grass and turtle grass in the lower laguna, near Brazos Santiago Pass, where salinities were lower due to the inflow of gulf water. Because of the prevailing high salinity, shoal grass and widgeon grass, the most salt-tolerant species, were the most abundant.

In the years prior to the opening of the waterway, ecologists believe, seagrass went through a "boom-and-bust" period. Seagrass, and the fish and wildlife that depended upon it, would flourish or retreat, depending upon the rainfall and the saltiness of the bay. Seagrass would spread after heavy rainfall, when salinity would go down to tolerable levels. Hurricanes, four of which hit the bay between 1916 and 1949, not only would dump huge amounts of fresh water but also would cut passes through Padre Island, allowing gulf water to enter. After these events of heavy rainfall, salinity would decline for several years. When periods of low rainfall returned and the passes filled in with sand, severely hypersaline levels would again dominate the system and seagrass would die out.

This storm-driven cycle of scarcity and abundance still occurs today within the Laguna Madre of Tamaulipas. The Mexican laguna, similar in configuration to its counterpart in Texas, does not have a shipping waterway cutting through it to alter its natural circulation. As with the pre-waterway Texas laguna, seagrass — mostly shoal grass — spreads when conditions are favorable and then retreats as drought and rising salinity return.

In 1967, Hurricane Beulah slammed into the northeastern coast of Mexico and South Texas with intense rainfall over the northern Mexican and Texas lagoons. Within months, the salinity had dropped to below 10 ppt in the Mexican laguna (and almost to 0 ppt in parts of the Texas laguna). Previously, during the drought of the 1950s, the Mexican lagoon had been little more than a shallow pool supporting mainly brine shrimp and some salt-tolerant insects, such as brine flies. Within two years of the hurricane, commercial landings of the spotted sea trout went from none to over a million pounds.

After the Intracoastal Waterway's completion, and the opening of Mansfield Pass through Padre Island, in 1962, more gulf water could enter the lower Laguna Madre than before. Prevailing southeastern winds blowing roughly parallel to the bay drove water currents north through the Land Cut into the upper lagoon and back out to the gulf through Corpus Christi Bay by way of Aransas Pass. This enhanced circulation resulted in lowered salinity across the bay system, as the less salty gulf water began to mix more thoroughly with bay water. The grass-killing hypersaline levels observed in the early decades of the century — often exceeding 100 ppt — were now an event of the past; average salinity plummeted to about half of the historic highs and became more stable. The stage was now set for the great changes to come.

"In the late 1950s, conditions in the Laguna Madre were just becoming favorable for seagrass," says Dr. Warren Pulich Jr., a leading seagrass expert and coastal ecologist with Texas State University–San Marcos. "Seagrass really began to spread in the 1970s." In the years after the waterway was first dredged, shoal grass began to spread, especially in the lower laguna, where salinities were much lower than in the more isolated upper laguna. By 1965 shoal grass covered almost the entire bay bottom of the lower laguna, except for small areas of manatee grass and turtle grass at the south end, near Port Isabel. But salinity eventually stabilized to lower levels that would allow manatee grass and turtle grass to grow. Within the range of salt tolerance of those two species, shoal grass lost its competitive advantage. Manatee grass and turtle grass have larger leaves and physically overshadow and outgrow shoal grass to the point of displacing it.

Much of the change in seagrass distribution in the Laguna Madre of Texas is governed by a process known as ecological succession, the change in species composition of a plant community over time. The image of seagrass as a meadow is more than just a convenient metaphor; succession applies as much to seagrass as it does to a pasture of golden aster and post oak in Fayette County.

Succession begins with a disturbed environment into which colonizing, or pioneer, species invade and establish themselves. Colonizers are the species that take first advantage of a disturbed environment and that grow quickly. Eventually, species that compete more successfully will displace the colonizers and become themselves the dominant species of the community. One successional community creates the conditions for the next. This cycle continues — successive stages of more competitive species displacing the previous ones — resulting in a climax community.

The natural world abounds in events that create disturbed habitats: floods, windstorms, earthquakes, drought, volcanic eruptions, hurricanes, landslides, and, perhaps the most common of all, fire. After a pasture burns from a lightning strike, for example, grasses and wildflowers — the colonizing species — will first appear. If the pasture does not burn again, woody shrubs and bushes move in, displacing the grass and flowers. Trees eventually invade, displacing the shrubs and bushes, and the plant community reaches its final successional stage, a climax forest.

In the Laguna Madre, shoal grass is a pioneer rooted submerged plant. It grows quickly and readily establishes itself throughout disturbed, un-vegetated bay bottom. Shoal grass generally grows in areas that are not favorable for any other seagrass. In the pre-waterway laguna, the disturbance event was the high salinity that killed or inhibited other grasses. As the lethal levels of salinity declined after the waterway was first dredged, shoal grass spread throughout the barren bay floor, first in the lower laguna and later in the upper laguna. But as salinity continued to drop, manatee grass and turtle grass, both larger and more aggressive, spread in succession throughout the lower bay at the expense of shoal grass. A huge reduction in winter use by the Redhead accompanied this shift in grass species. Mid-winter surveys conducted by the Texas Parks and Wildlife Department between 1956 and 1966 indicated that 70 to 90 percent of the Texas coastal population of Redheads used the lower bay. Today that percentage has declined, but recent surveys by waterfowl researchers at Texas A&M University–Kingsville have shown that Redheads are still more abundant in the lower laguna.

Throughout the thornscrub and seacoast bluestem of the adjacent King and Kenedy ranches lie hundreds of small ponds and depressions that have been carved out of the sandy soil by the everlasting wind. Some ponds are small potholes, known as *copitas* — "little cups" — or they may be elongated swales; others are oxbows, remnants of the ancient course of the Rio Grande. They capture and hold rainfall and serve as the main source of fresh water for the Redhead. "The coastal ponds are very important for the Redhead as a source of fresh drinking water and as a resting place," says Dr. Ballard.

The Redhead's abrupt change in habitat from freshwater prairie wetlands to a salty coastal bay each fall creates a sudden demand for freestanding fresh water. As the bird feeds on shoal grass, it ingests excessive salt that must be dealt with to avoid salt stress. The Redhead excretes this excess salt by means of the salt gland, a specialized structure near the eye found in numerous species of marine birds and reptiles. During the birds' breeding season the salt gland shrinks from disuse; it takes several months for the gland to become acclimated. The bird manages its toxic jolts of salt by drinking fresh water.

These inland freshwater ponds are used all winter and are recognized as an essential component of the Redhead's coastal habitat that needs to be protected. Many have been destroyed by land levelling for agriculture or other development, with almost half of the seasonal ponds adjacent to the lower bay having been lost. "The loss of coastal freshwater ponds, especially those within close proximity to the abundant shoal grass resources of the Laguna Madre, is among the biggest long-term threats to the Redhead," says Dr. J. Dale James, Director of Conservation Planning for the southern region of Ducks Unlimited.

In 2012, a team of coastal science experts known as the Basin and Bay Expert Science Team (BBEST) published and submitted to the Texas Commission on Environmental Quality a comprehensive report that revealed a disturbing trend within the lower Laguna Madre. The report, entitled "Environmental Flows Recommendations Report," presents a detailed evaluation of the watersheds of the lower bay and the Rio Grande estuary and of the effects of freshwater inflows on the bay's ecology. Technical specialists from the Texas Water Development Board and the Texas Parks and Wildlife Department also assisted the expert team.

The report stated that, up to the early 1990s, the lower bay was in generally sound ecological condition, despite the successional changes in species that occurred since the dredging of the Gulf Intracoastal Waterway. Until then the lower bay had the clear water for which it was long renowned; dense meadows covered the bay bottom and grass could grow at depths of six feet. "But water quality has declined in the lower laguna over the past 20 to 30 years," says Dr. Pulich, one of the participating scientists in the BBEST study. As a result, total seagrass acreage has undergone a decrease of over 20 percent within the lower bay. Shoal grass has suffered an especially severe reduction, with a decline in acreage of around 60 percent from its peak in the mid-1960s.

Decreased water quality and clarity may be the likely cause of the decline in seagrass observed in the lower bay. As the human population of the Brownsville and Lower Rio Grande Valley has grown, agricultural and municipal drainage through the Arroyo Colorado in the lower bay has increased. The BBEST team concluded that the flow through the arroyo and other channels harms the lower laguna in two ways.

First, during periods of heavy rainfall, large volumes of fresh water enter the bay, creating areas of low salinity that stress seagrass. Second, during periods of low rainfall, the agricultural and municipal drainage dominate the flow, carrying heavy loads of nutrients such as nitrogen and phosphorous, which stimulate the growth of microscopic organisms and seaweed, all of which diminish water clarity. The BBEST team has made recommendations for restricting freshwater inflow to mitigate the adverse effects of excess nutrients.

Maintenance dredging of the Gulf Intracoastal Waterway has long been recognized as another threat to seagrass. The construction of this channel did more than just alter the salinity regime of the bay. Seagrass beds were dredged up along the western edge of both bays by the initial channelization. Routine dredging operations subsequently have been documented to cause seagrass loss because they disperse light-reducing sediment in the water column. The 12-foot-deep channel must be dredged every two years to maintain navigability; state and federal agencies have conducted numerous studies to identify alternate disposal areas and beneficial uses of dredge spoil. Regardless of location, however, submerged disposal mounds serve as a continuing source of suspended sediment as the wind blows over the bay. Because of the damage to seagrass, the location of dredge spoil has become contentious over the years. The Texas Department of Transportation has coyly noted that "[v]arious groups have objected to some of the traditional locations where dredged materials are placed." The agency, however, still insists on dumping dredge spoil in the Laguna Madre despite the scientifically based objections.

Chesapeake Bay offers a cautionary tale for the Redhead. That 2,500-square-mile estuary, three times the size of the Texas Laguna Madre, meanders throughout the shores of Maryland and Virginia, formed in part by some of the most storied rivers in American history: the York, the James, the Potomac, the Rappahannock. From the head of the bay, at Susquehanna Flats, to 200 miles south at its opening to the Atlantic, Chesapeake Bay had in the first half of the twentieth century luxuriant communities of seagrass and other edible plants growing in the clear, shallow waters. This bounty supported great flocks of Redheads in winter, reaching at one point 120,000 birds during the 1950s, a period when all nesting prairie waterfowl were abundant.

The Redhead would find winter forage — eel grass, widgeon grass, sago pondweed, and many others, including a grass known as redhead grass — throughout the bay and its tributaries; shoal grass, a subtropical species, did not reach as far north as Chesapeake Bay. But then, year after year, decade after decade, with increasing rapidity, seagrass began to disappear. First in the upper bay and the upriver environs, then extending to the lower bay and its tributaries, seagrass meadows died out to the point that, by the 1980s, entire river systems had no submerged, rooted plant life. Ecologists would describe this collapse as unprecedented in the bay's recent history. The nation's largest estuary had come under enormous environmental stress.

As the seagrass vanished, so did the Redhead. Within twenty years of its mid-century peak, the winter population plummeted, in some years to fewer than a thousand birds. Although the duck and seagrass have recovered somewhat in recent years, "the Redhead today is a shadow of what it used to be in Chesapeake Bay," says Larry Hindman, former Waterfowl Project Manager with the Maryland Department of Natural Resources.

Drought and degraded water quality from pollution are believed to have been the cause, at least in part, of this calamitous loss of aquatic plants. Tropical Storm Agnes, which struck the area in 1972, was especially destructive, as it poured enormous amounts of fresh water into the system, uprooted seagrass beds, and dumped tons of sunlight-killing sediment into the water.

As the winter population of Redheads in Chesapeake Bay dwindled, that in North Carolina increased. Areas such as Pamlico Sound and Core Sound saw an eight-fold increase, to over 40,000 birds, by the early 1980s. Then, as North Carolina's seagrass died out, the Redhead departed and continued its southern exodus down the Atlantic Flyway to Florida. As Redhead numbers in Chesapeake Bay and North Carolina fell, they rose along the Florida Gulf Coast.

The wanderings of the Redhead in search of winter food appear to have been a recurring theme in the story of this unique diving duck. The ornithologist John C. Phillips observed this phenomenon in the early 1900s when he wrote, in his four-volume monograph *A Natural History of the Ducks*, that the Redhead "is so particular about its food that it is extremely irregular in its appearance . . . that in regions where it is usually seen in thousands it may almost disappear for a term of years." Philips also notes that, as a result of this searching for food, " . . . wild stories of its increase or decrease are circulated from time to time, and it becomes . . . very difficult to make a fair guess at its actual status in any one place."

A reasonable question to ask is where the Redheads spent the winter months before 1949 — that is, where did the bird go during periods when the Laguna Madre was largely devoid of seagrass, because of the extreme salinity that prevailed before the construction of the Gulf Intracoastal Waterway. The most likely answer is that the Redhead was simply more widespread than it is today — over the Texas Coast and northeastern Mexico, in Louisiana and Alabama, California, and Florida, along the Mid-Atlantic Coast, wherever the bird could find shoal grass or other edible species. This wider distribution of the Redhead along the Texas Coast was alluded to by Phillips in his *Natural History*, published in 1925, where he writes that "from various reports, [the Redhead] seems to have a tremendous center of abundance all along the lagoon region of southeastern Texas away down to Cameron County."

Reliable population estimates from this period are not available for the Redhead, or for any other waterfowl. Since the earliest days of North American settlement, explorers and naturalists observed and studied the continent's bird life. The U.S. Fish and Wildlife Service, however, did not conduct formal, annual bird surveys yielding numerical estimates of populations until 1947. When the laguna would undergo its periodic "boom" cycle, after a major storm or rainy season that reduced salinity for several years, the Redhead most likely utilized the bay to a greater extent. When high salinity returned, the seagrass died out and the Redhead moved elsewhere.

The Redhead demands much of the North American landscape. Its need for a variety of habitats — freshwater prairie wetlands, alkali marshes, large, remote lakes for post-breeding molting, migratory resting places with adequate food and water, a rare coastal marine embayment for a winter refuge — and its relentless dependence upon a scarce and diminishing winter food place *Aythya americana* among the more challenging of waterfowl for long-term stewardship.

Waterfowl have endured, since the early nineteenth century, commercial slaughter, unregulated hunting, historic drought, poisoning from lead shot, mass mortality from disease, and governmental agricultural policies that destroy habitat. "It is amazing how waterfowl have persisted despite everything we have thrown at them," says Kevin Kraai, the Waterfowl Program Leader with the Texas Parks and Wildlife Department. "These are very resilient, very hardy birds." With the protection of federal and state game laws and the support of public and private conservation and management programs, duck and geese have, for the most part, continued to thrive. The Redhead, in fact, has enjoyed populations well above its long-term average in recent years due to good rainy seasons in the breeding range, according to surveys conducted by the U.S. Fish and Wildlife Service. But, as Kraai also says, a "perfect storm" looms over the world of waterfowl, not just in Texas but across the continent, borne of the ever-increasing loss and degradation of vital landscape and the threats of drought, disease, and climate change.

Like those of other species of waterfowl, Redhead population numbers have varied widely over the years, as rain, snowmelt, and drought affect nesting habitat and food supply from year to year. North American ducks did well in the 1950s, with a total breeding population of around forty million, but took a serious plunge in both the early 1960s and again in the early 1990s to about half of the 1950s estimate; the intervening years saw a good recovery to previous levels. A 2018 estimate of breeding waterfowl in Canada and the Midwest places the population at approximately forty-five million birds. Several years of adequate rainfall throughout Prairie Canada and the north-central United States, where most of the North American ducks nest and raise young, have enriched the grasslands and pothole wetlands and have sustained this bounty.

Duck populations, however, can crash with startling swiftness. Breeding range conditions can quickly deteriorate, with population data for the past fifty years showing that total duck populations have dropped several times by almost half in less than a decade. Many individual species show reductions by more than 20 percent in just one year; the Green-winged Teal, for example, fell by 60 percent over one 12-month period. These population swings result from periods of rain and drought, during which the number of seasonal ponds — prairie potholes with standing water — rises and falls by the millions. The recent fortunes of the Redhead vividly demonstrate these extreme changes.

Estimates of the Redhead's breeding population have ranged from three hundred thousand birds in the early 1960s to almost 1.4 million birds by 2011, a figure well above its long-term average. In 2017, according to the annual Mid-winter Waterfowl Survey conducted by the Texas Parks and Wildlife Department, the statewide population of Redheads stood at over 850,000. But the following year it fell to around 86,000. For just the Texas Coast, the 2018 estimate was only 43,000, compared to 690,000 birds the previous year.

For more than a century, the Prairie Pothole Region, the cradle of the continent's waterfowl, has been under assault. The early settlers, who long ago discovered the richness of the prairie landscape for growing crops, began to drain the wetlands and plow the grasslands in a slow, steady destruction that has continued unabated to the present day. Year after year, tens of thousands of acres of small ponds and their embracing grasslands are forever wiped out. A 1988 report by the Secretary of the Interior stated that, of the original twenty million acres of wetlands in the US portion of the pothole region, seven million acres remain. Since then losses have continued. Minnesota and North Dakota have lost almost half of their original wetlands, and South Dakota and Montana about a third. Today Iowa has one percent of its original prairie potholes. The exploitation of wetlands and grasslands has only intensified over the past two decades under the ever-increasing demands for food and energy.

The pressure to consume the Prairie Pothole Region comes from many directions: rising values for agricultural lands, which have doubled over the last ten years; increasing agricultural commodity prices, especially for corn and soybeans; government mandates for biofuels; the extraction of oil and natural gas; the expansion of wind energy farms (as of 2010 there were 3,759 wind turbines in the US prairie pothole area); decreasing budgets for governmental and private conservation programs; US farm programs that encourage the tilling of lower quality prairieland by offering financial assurance; enhanced technologies for improving yields on marginal land and for draining ponds; and displaced prairieland from urban expansion.

Moreover, government programs that benefit wetlands are being scaled back. The Conservation Reserve Program, for example, a component of the Farm Bill that encourages private landowners to conserve wetlands, is reducing the amount of eligible acreage; nationwide, almost three million acres have been removed from the program, included over a million in just the Prairie Pothole Region. Moreover, several legal rulings — such as the 2006 Supreme Court ruling *Rapanos v. United States* — have resulted in almost all the wetlands in the pothole region losing their legal protection under the Clean Water Act. In addition, wetlands have very little regulatory protection in Canada. In short, the economic incentive for plowing up the Prairie Pothole Region is enormous and very little stands in the way to prevent it.

This is more than a problem for only the Redhead, or even for only waterfowl. A host of our native birds depends upon the region: migratory shorebirds such as curlews and plovers, grassland birds such as larkspurs, bobolinks, and sparrows — some 200 species in all. Ducks Unlimited, the nation's leading waterfowl conservation group, and the North American Waterfowl Management Plan, produced by the governments of Mexico, Canada, and the US, have identified the Prairie Pothole Region as their top priority for conservation programs.

But there is yet another disturbing sign for waterfowl, one that is perhaps not immediately apparent — and is even somewhat contradictory — but is insidious nevertheless: the continuing decline of interest in duck hunting. In the United States, the number of active waterfowl hunters has declined from two million in 1970 to less than a million in 2015. Duck stamp sales have fallen accordingly. Duck hunting in Canada has dropped by more than half. The nation's rich heritage of waterfowling no longer occupies the place of importance in modern life that it once did. "We don't fully understand all the reasons for the decline in hunting, but it is probably a combination of factors," says Todd Merendino, Manager of Conservation Programs for Texas Ducks Unlimited. "People are busy, leases are expensive and less available, and duck hunting is more physically demanding than hunting deer or quail." But the basis for the concern over the decline is well-defined: economics.

As the *North American Waterfowl Management Plan* states: "In the United States, waterfowl conservation has ridden, by and large, on the coattails of waterfowl hunters, who have been the strongest advocates for conservation policies and large financial contributors to waterfowl conservation." Since its inception with the 1934 Migratory Bird Hunting and Conservation Stamp Act (the "Duck Stamp Act"), the duck stamp program — which featured the Redhead on the stamps in 1946, 1960, 1987, and 2004 — has generated over $960 million dollars for wetlands and waterfowl conservation. A decrease in this important revenue stream is yet another strike against preservation of vital duck habitat.

The management plan clearly identifies the heart of the problem as that "waterfowl populations are facing unprecedented threats, and current levels of conservation are unsustainable without reversing trends of hunter decline and garnering more support from a broader constituency." As former Secretary of the Interior Ken Salazar has rightly said: "The biggest single threat to conservation in America is the growing disconnect of our people with the outdoors."

Waterfowl and their habitats have benefitted enormously, and continue to benefit, from efforts of private citizens, private conservation groups, and governmental agencies; over the years they have preserved millions of acres of landscape that have well served the nation's waterfowl. Nothing is more important to maintaining waterfowl populations than good habitat for nesting, migration, and wintering. If a greatly diminished Prairie Pothole Region is hit with a drought that lasts several years — the 1950s drought in Texas lasted seven years — and the weakened waterfowl populations have only a coast degraded by human activity and climate change to which to retreat for the winter, the "perfect storm" will have come to pass. An era of uncertainty lies on the horizon for North American waterfowl.

The future of the Redhead depends upon the future of the Laguna Madre. If, two or three or five decades from now, the slow changes that have been moving across the bay lead to the eventual disappearance of shoal grass, the Redhead will need to seek its winter refuge elsewhere as it did in the era before the waterway. But, unlike during that time, the Redhead will have fewer places to go. If the vital natural endowment that the upper Laguna Madre has long enjoyed holds true — the unspoiled expanses of ranches and barrier islands that embrace it — shoal grass may survive the coming decades. "As long as the King Ranch stays undeveloped and North Padre Island remains intact," says Dr. Kim Withers, a marine biologist at Texas A&M University–Corpus Christi, "the upper Laguna Madre should be too salty for manatee grass and turtle grass to continue spreading. Shoal grass should persist in substantial amounts." If the damage to water quality in the lower laguna can be abated, seagrass may reverse its steady decline. If not, the resilience long displayed by the Redhead — the duck that depends upon the Texas Coast more than any other North American migratory waterfowl — will yet again be called upon for its survival.

LINDA M. FELTNER

CHAPTER 5

Bolivar Days

Birds of Spring, Birds of Fall

WHEN HURRICANE IKE RAKED the fury of its 110-mile-an-hour
winds across Bolivar Peninsula in September 2008, it inflicted enormous damage upon the 27-mile long, low-lying strip of sand along the
upper Texas Coast. Located east of Galveston, Bolivar received the
full blast of Ike's power as the eye of the storm passed over the island
city. The small communities along Highway 87 — Port Bolivar, Crystal
Beach, Caplen, Gilchrist — fell to the wind, waves, and twenty-foot
surging wall of water, their beach homes, convenience stores, and clap-
board hamburger and beer joints shattered into piles of woody rub-
ble. The onslaught tore up beaches and dunes, drove sand far inland,
fractured stretches of the highway, ripped up trees, and swept away in
minutes the crests of 2,000-year-old ridges, formed by the ancient ebb
and flow of the gulf, that ran down the spine of the peninsula. At one
point Bolivar's landscape of grass, wildflowers, and mottes of oak and
hackberry trees lay beneath ten feet of water from Ike's immense storm
surge, the largest on record for the Texas Coast since Hurricane Carla,
in 1961. In Gilchrist, the community surrounding the popular fishing
spot Rollover Pass, all but two buildings were destroyed. "Gilchrist
was wiped off the map," says Claud Kahla, a lifelong resident of nearby
High Island who owned a hardware store in the devastated communi-

ty. The cost of Hurricane Ike stood at $30 billion dollars, at the time making the 275-mile-wide, Category 2 tropical cyclone the third costliest storm ever to strike the US mainland.

Ike's legacy of destruction, however, went beyond the loss of human life — estimated at seventy-four deaths in Texas — and property of the peninsula's 2,400 residents. The damage done to two areas on Bolivar — Bolivar Flats, at the south end, and High Island, at the opposite end — threatened not only local birdlife but also birds that, when Ike struck the coast, were thousands of miles away. Bolivar Flats and High Island are two of North America's most important coastal havens for mid-continental migratory shorebirds and songbirds, as two great annual events in the world of Gulf Coast bird life occur there: the arrival of wintering shorebirds at Bolivar Flats, from mid-summer to fall, and the spring migration of songbirds at High Island. Sandpipers and plovers raising young in the far northern reaches of Canada and Alaska and warblers and tanagers flitting around the jungles of Central and South America — along with many other bird species — at some point in their lives will need the food and shelter provided on Bolivar Peninsula.

Bolivar Flats is a 1,200-acre piece of shoreline on the southern end of Bolivar Peninsula, the northernmost link in the Texas chain of coastal barriers and the northern boundary to the entrance of Galveston Bay. To the untrained eye the flats offer little in scenic allure: a quiet expanse of mud, sand, marsh, and uplands where the gentle twice-daily tides and everlasting sea breeze constantly move the water about, creating shallow ponds, mudflats, tide pools, and tidal channels. A community of marine creatures — snails, worms, shrimp, crabs, beach fleas, *Donax* clams — lives beneath the surface and among the green, dense gatherings of *Spartina*.

Owned and maintained by Houston Audubon, Bolivar Flats — formally known as the Bolivar Flats Shorebird Sanctuary — has been given the rare distinction of a Globally Important Bird Area by the Western Hemisphere Shorebird Reserve Network. This designation is based upon either an annual shorebird population of more than 100,000 birds observed per year or serving as habitat for more than 10 percent of the continental population of a species. This network, a consortium of scientists and conservationists from fourteen nations, has identified ninety-two such shorebird areas from the Arctic to Tierra del Fuego.

On the opposite end of the peninsula sits High Island, a town of around 500 inhabitants perched upon a salt dome that, at thirty-eight feet above mean sea level, is the highest elevation on the Gulf Coast between Mobile, Alabama, and the Yucatan Peninsula. From miles away, the town's gently sloping rise comes into view, blanketed with dense oak forests, like an island in a quiet green sea. The High Island bird sanctuaries comprise four separate parcels totalling 255 acres, all of which are also owned and maintained by Houston Audubon: the Smith Oaks Sanctuary, the largest at 177 acres; the 60-acre Boy Scout Woods (so-called after the former scout camp on the site, but officially named the Louis B. Smith Bird Sanctuary); Eubanks Woods (named after former Houston Audubon president Ted L. Eubanks Jr., who played a pivotal role in establishing the sanctuaries); and the S. E. Gast Red Bay Sanctuary. A rookery of egrets and spoonbills, as well as ponds, wetlands, and patches of coastal prairie, lie throughout the woodlands.

Each spring as migrating songbirds come north across the gulf from the tropics, the woods of High Island come alive with flutter- ing colors of warblers, tanagers, orioles, vireos, kingbirds, catbirds, grosbeaks, thrushes, and buntings as they land, exhausted and starv- ing, into the welcoming canopy of the sanctuaries. The annual spring arrival, as with the winter arrival of the great cranes at Aransas, is big business: around 6,000 birders a year visit the sanctuary just for spring migration (April and May), many of them taking advantage of the bleachers that are provided as if for watching a high school football game. Houston Audubon estimates around 10,000 birders visit High Island annually.

Along with the Aransas National Wildlife Refuge, Bolivar Flats and High Island are the most internationally renowned of Texas's birding locales. A veteran Texas birder and High Island resident tells the story of encountering an Englishman on a nature tour in California. Thinking that the Briton would not know the location of the small coastal com- munity, he told him that he lived "near Houston." The Englishman was apparently not familiar with the geography of the nation's fourth-largest city but he immediately recognized the name of High Island.

At Bolivar Flats, Ike wiped out the dunes and scalped the beach of four feet of sand, exposing in places the mud laid down during the last Ice Age, dumping tons of sand inland and pushing the coastline landward about 200 feet. The invertebrate community that served as a food source for migrating shorebirds disappeared.

At High Island, Ike's storm surge caused flooding up the slopes of the salt dome, but most of the community and the woodland sanctuaries were spared inundation. The town also had to contend the year before with Hurricane Humberto, a strong Category 1 storm that struck the upper Texas Coast in September 2007. Humberto's powerful winds knocked over and tore up huge trees, many of which stayed rooted but without their majestic, cathedral-like arches. After Ike had subsided, the questions that many had were: Will Bolivar Flats and the woodlands of High Island survive and recover? Will the shorebirds and songbirds ever return?

"Bolivar Flats are special for artificial reasons," says Dr. Richard Gibbons, Conservation Director of Houston Audubon. One of these artificial reasons is the five-mile-long red granite jetty that reaches into the gulf from Port Bolivar, located near the south end of the peninsula. "The flats are produced by the large eddies of longshore currents created by the jetty," he says. "As the gulf water slows and circulates back around, mud, sand, and organisms fall out and a fresh supply comes in every day with the tides. The result is a large intact marsh with a dense concentration of food all year for shorebirds."

Before the emergence of Houston as a major port with the opening of the Houston Ship Channel, in 1914, Galveston in the late-nineteenth century was the busiest port on the Texas Coast. Prior to the original construction of the jetty, which began in 1874, sandbars formed at the mouth of the bay and blocked the entrance of large ships into the port. Gulf water entering Galveston Bay with the tides deposited huge amounts of sand — from the Mississippi and Sabine rivers — at the mouth of the bay, creating the sandbars and giving access to the inner harbor only to shallow-draft shipping. Larger ships would have to wait offshore while their cargo — cotton, grain, beef, farm machinery, and much else — had to be handled by smaller vessels that could go over the sandbar. This cumbersome situation was expensive and inefficient, and the city authorities soon realized their competitiveness in the growing economy would decline if the port could not accommodate deep-draft shipping.

The first attempts at constructing jetties with timber, stone rip-rap, and large sandbags proved ineffective, since they eventually washed away or were eaten by marine organisms. A more long-lasting structure was needed, one that could withstand the constant punishment of the gulf. The problem of sand build-up at the mouth of the harbor persisted. The solution emerged from Washington, DC.

An 1890 *New York Times* article reported on the competition for Congressional appropriations for a deep-water port west of the Mississippi; Sabine Pass, Corpus Christi, Brazos Santiago, and Aransas Pass, as well as Galveston, all had their backslapping lobbyists prowling the halls of Congress. "Galveston, advantageously situated with a magnificent bay at her back," the article stated, "is seriously handicapped by her inaccessibility to vessels of great carrying capacity . . . . Delays, vexatious and costly, naturally attend the present system at Galveston." Galveston was awarded the six-million-dollar contract and the U.S. Army Corps of Engineers soon embarked upon one of its most cherished pursuits: impeding the flow of water.

Granite jetties at the mouth of the bay were proposed as a permanent solution to the constant need to dredge the bay bottom to keep the shipping lanes open. By 1898, the Corps had completed both the North Jetty, which today extends almost five miles into the gulf from the south end of Bolivar Peninsula, and the two-mile-long South Jetty, which extends from the north end of Galveston Island. The giant hunks of red granite that cover each jetty, some up to ten tons each, came from the granite quarry near Marble Falls, which provided the stone for the state capitol. "It is admirably suited for this purpose," says a 1917 monograph on the state's granite resources by the University of Texas, "being heavy and highly resistant to the corrosive action of sea water."

The entrance to Galveston Bay was now protected by the two jetties, with the sand carried by the westerly longshore currents along Bolivar being intercepted by the North Jetty. The sand swirled around and settled out, creating Bolivar Flats, a ragged bulge in the coastline at the tip of the peninsula.

Shorebirds begin to increase in numbers on Bolivar Flats toward the middle of summer as they depart their nesting grounds in Canada and the northern US and head south for the winter. From October to May, the season of their heaviest use of the flats, birds reach their peak numbers, averaging more than 10,000 birds a day.

More than 320 species of birds have been recorded on the flats, including most of the fifty-three species of North American shorebirds from the Charadriiformes, the taxonomic order that encompasses the shorebirds, gulls, auks, and terns. Sanderlings and sandpipers dash about the surf-washed beach, pecking at food at the surface with their short beaks; rails and Seaside Sparrows move throughout the *Spartina*; herons and egrets hunt for fish in shallow ponds, inlets and tidal pools; curlews with their long beaks probe for food deeper in the marsh soil; and the mudflats, flooded and then drained and flooded again with the tides, host the feeding avocets, dowitchers, stilts, godwits, and plovers. Roseate Spoonbills, pelicans (both brown and white), hawks, ibises, waterfowl, herons, sparrows, and egrets (including the threatened Reddish Egret) will make the flats their home for a part of the year. In spring and summer, Least Terns and Wilson's Plovers nest on the beach and on dry pans and Clapper Rails and Seaside Sparrows nest in the marsh grass. The endangered Piping Plover finds a home here, as does the threatened Snowy Plover. Gulls, terns, and skimmers keep company with the shorebirds for hours on the mudflats but fly off to feed in the gulf.

Some species stay briefly before moving south while others remain longer on the flats; some are most numerous in the summer, others more abundant in winter and spring. Other species may eventually travel south to the tropics, or they may work their way along the Texas Coast to other shorebird areas, such as the Laguna Madre of Texas and Tamaulipas, the South Texas Salt Lakes, the Anahuac National Wildlife Refuge, and the Texas Mid-coast National Wildlife Refuge Complex. (These locations have also been recognized as shorebird areas of international importance by the Western Hemisphere Shorebird Reserve Network, the only other locations on the entire Gulf Coast to have received this distinction.)

The American Avocet, the large black-and-white shorebird with a rust-colored neck and rakishly upturned bill, dominates the scene at Bolivar Flats. "The most numerous species here is the avocet," says Winnie Burkett, a former sanctuaries manager for Houston Audubon who played a key role in the sanctuary's development and who owns a beach house overlooking Bolivar Flats and the North Jetty. "We will commonly see 5,000 avocets a day in winter and in spring, sometimes as many as 10,000 to 12,000." More than 17,000 avocets were observed in one day in April 2014. Regardless of the season, this unique area always has a wealth of bird life. "There is not a day of the year," says nature tour consultant Ted Eubanks, "that Bolivar Flats is not interesting. There is no better place on the Texas Coast to watch shorebirds."

Bolivar Flats had been known for years as an outstanding birdwatching area. One day in late 1984, however, a concerned citizen by the name of Stennie Meadours appeared before the Galveston County Commissioners Court with a simple but controversial request: declare 900 acres of beach, marsh, and mudflats at the tip of the peninsula as a shorebird sanctuary and restrict vehicular traffic.

Meadours, who at the time served on the board of Houston Audubon, had been birdwatching for only a few years but had soon become captivated by Bolivar Flats. She would leave her home in La Porte, a small town between Houston and Galveston, and routinely visit the flats to observe the enormous flocks of shorebirds. But Meadours was also appalled by what else she saw. "There would be thousands of birds roosting and feeding on the flats," she says, "and kids on 4-wheelers would drive right through them, sending the birds circling above until the vehicles left." People would drive cars at will all over the beach and flats, crushing eggs and chasing off birds. Meadours vowed to do something about this.

Three years prior to her request before the court, Meadours had quietly instituted a campaign to establish Bolivar Flats as a shorebird sanctuary. She first brought the need for protection of nesting and roosting birds on Bolivar Flats to a meeting of the Houston Audubon Board, which promptly voted to authorize her to act on the society's behalf to pursue a Galveston County ordinance to restrict vehicular traffic on Bolivar Flats.

Part of that campaign, which included obtaining newspaper coverage in Houston, Galveston, and Beaumont newspapers, as well as endorsements from local nature groups, was a monthly, yearlong bird survey of the flats. "The protection of the flats is a worthy cause," the *Galveston Daily News* was to declare in an editorial, "and one that [the people] of Galveston County . . . should support."

Armed with extensive bird data and supported by Houston Audubon, Meadours stated her case before the Commissioners Court in November 1984. The heart of Meadours's proposal was simply to legally prohibit cars, motorcycles, and other vehicles from the beach at the flats, thereby protecting feeding, roosting, and nesting areas. People would still be able to walk and fish around the flats. The area in question was public land (the area between the mean high tide line and the gulf), and it was determined that Galveston County had the right, if it wished, to restrict vehicular traffic. This apparent threat to the long-revered right to drive on Texas beaches was immediately opposed by people who feared the loss of fishing and other uses of the area.

By the time the public hearing on the matter was scheduled, in February 1985, opponents of the proposal had assembled a petition with more than 900 signatures, primarily of Bolivar residents. One objection was that senior citizens would have to walk too far to get to the flats. "Any restrictions there," said the leader of the opposition, "are unnecessary and unwarranted." The local concern about this issue was widespread enough to cause the Commissioners Court to delay a decision, pending further review.

Stennie Meadours and the birds, however, eventually won the battle. On July 1, 1985, Galveston County Commissioners Court, led by now-retired County Judge Ray Holbrook, voted to restrict vehicles from Bolivar Flats. "I wasn't a birdwatcher myself," Judge Holbrook says, "but I knew Bolivar Flats was an important area for birds." At the public hearing, Meadours presented the court with a photograph of a tire track one inch from the egg of a Least Tern. "I think that photo was very effective in convincing the court that vehicular restrictions were needed," Meadours says. She remembers that when she arrived at the public meeting, her vehicle was the only car; all others were pickup trucks.

Eventually, and despite continuing local opposition, Houston Audubon, through the efforts of society members Gretchen Mueller and Ted Eubanks, was granted a lease from the Texas General Land

Office for the protection and management of 550 additional acres of marsh, beach, and uplands adjacent to the beach. In April 1992, the Texas Parks and Wildlife Department formally dedicated Bolivar Flats as an official shorebird sanctuary, the first of its kind on the Texas Coast. The following decade saw the purchase of additional parcels of marsh, mudflats, and uplands from private landowners through the efforts of Winnie Burkett and others, to reach the sanctuary's present size of 1,206 acres. "I planted the seed," Meadours says, "but Winnie Burkett and Houston Audubon grew an orchard."

A century before Bolivar Flats came under legal protection another metaphorical orchard had been planted on the opposite end of the peninsula. This was in the area known during the days of the Republic of Texas as the "High Islands." As it happens so often within the story of Texas coastal bird life, private property owners played a decisive role in the preservation of bird habitat.

As described in Houston Audubon's history, the bird refuges of High Island had their origin with the nineteenth-century settlers who planted trees and created woodlands and their descendants who enhanced and preserved them. Like the surrounding landscape, the area on and around the salt dome was then covered with prairie grass and wetlands. In 1874, John Brown acquired the property; according to family lore, he loved oak trees and planted them not only on his property but also throughout High Island. A live oak tree with a diameter of more than seventeen feet and a canopy almost half the length of a football field has a plaque on it proclaiming it the "John Brown Oak." Brown's daughter Charlotte eventually inherited his wooded property and she and her husband George E. Smith planted vines, citrus, flowers, and more trees. By the late 1950s, after three generations of descendants of John Brown protecting the woodlands, word got out among the small community of birdwatchers that the oak woodlands of High Island were indeed a special place.

As High Island grew in popularity with birders throughout the 1960s, conservation groups began to look upon the oak woods as wildlife habitat that deserved long-term preservation. In 1970, the Texas Chapter of The Nature Conservancy made the first effort to negotiate for a possible acquisition. While this attempt was unsuccessful, the modern-day landowners not only continued to honor their family's desire to preserve the land but also declared the woods a private sanctuary open to birders, the first "Smith Oaks Bird Sanctuary."

Houston Audubon eventually purchased the eleven-acre Smith homestead in 1987. Purchase of additional tracts by the society and a donation of 110 acres by Amoco finally brought the Smith Oaks sanctuary to its present size.

The Louis B. Smith Bird Sanctuary originated when Mr. Smith (no relation to the Smith Oaks family) moved to the area in the mid-1940s and worked in the local oil fields. By the 1970s he decided to sell the richly wooded site but wanted to preserve it as a bird sanctuary. He asked Ted Eubanks, who had been birding High Island for years, about finding a buyer. Eubanks, who owned three wooded tracts adjacent to Mr. Smith's property that he later donated, first approached the Houston Outdoor Nature Club about acquiring the land; this proposal was emphatically declined. Eubanks then asked Houston Audubon about buying the land and the society agreed. However, the group immediately faced the challenge of raising funds for the most expensive project it had ever undertaken.

Not everyone within the society agreed with the acquisition of Mr. Smith's woods, and there was much contentious debate. Since the Board of Directors was to be responsible for the purchase, several board members resigned over concerns of financial liability. In the end, vision won over risk and the project moved forward.

A campaign of auctions, promotions, and publicity soon followed as volunteers and conservationists set out to make this effort — the society's first coastal bird habitat — a success. Financial support came from hundreds of individuals and nature groups from all over the country. The Louis B. Smith Bird Sanctuary was eventually dedicated by Houston Audubon in 1982.

The following years saw steady growth of the High Island sanctuary system. After the establishment of the early parcels, other acquisitions took place with the support of individuals, organizations, and industry: another large donation of land by Amoco, grants from the U.S. Fish and Wildlife Service and Phillips Petroleum, and substantial private donations. Houston Audubon's goal of preserving this special home for birds had become a reality. Today Houston Audubon is one of the largest landowners among US Audubon chapters.

Migration evolved early in the life history of birds so that they could take advantage of seasonal changes in the availability of food and habitat for nesting. The ability to depart a winter-hardened region where insects, frogs, fish, plants, and other food sources would be scarce, and to return the following spring or summer when food would be abundant, allowed a greater likelihood of survival and raising young. With their ability to fly, intense metabolism, highly developed nervous systems, and powerful vision, birds are well adapted to periodic, large-scale movements in their annual cycle.

Migration is a complex phenomenon that is hazardous and costly in terms of both mortality and energy requirements. Despite these drawbacks, migration occurs widely throughout the world of birds; approximately 80 percent of North American birds undertake some form of this annual journey. The breeding season for most North American birds is in the spring and summer throughout the US and Canada. They then migrate south in the late summer or fall to warmer climates and return to their breeding range the following spring and summer.

Migratory distances vary greatly among birds, and some species take a different route in the fall than in the spring. Some migrants stay entirely within the US and Canada, moving between a northern breeding area and a southern range north of Mexico. But most North American land birds go south of the United States, to Mexico, Central America, and the Caribbean. Others migrate to northern South America, in areas like coastal Ecuador and Venezuela, while some go as far as Amazonia. Some species, predominantly shorebirds, travel even farther, to the Argentine pampas and beyond. Each autumn more than a billion birds of more than 200 species fly south. Then, in spring, they return to their summer breeding grounds throughout Canada and Alaska and the eastern forests and Great Plains of the United States. Birds that nest in the US and Canada and spend the winter in the tropics are known as Neotropical migrants.

The great arc of landmass that embraces the Gulf of Mexico — the coasts of western Florida, Alabama, Mississippi, Louisiana, Texas, and

Tamaulipas to the Yucatan Peninsula — contains some of North America's most valuable landscape for birds in migration. When migrants cross an enormous natural barrier such as the gulf, many arrive on land starved and emaciated from the intense burning of energy-rich fat reserves during the long flight. Studies of mist-net-captured spring migrant songbirds in coastal Louisiana and Mississippi coming north from Central or South America suggest that individuals in the poorest body condition rely most heavily on the coastal habitats that they can reach. They stop at first landfall and stay longer, while birds in better condition may keep going inland or stay at the coastal stopover a shorter time. The location of a coastal refuge such as at High Island, rather than at a woodland twenty or thirty miles farther inland, can mean the difference between survival and death for these severely stressed birds.

A field guide map of a migrant's range, with its neat patches of color depicting the summer range in this location and the wintering range in that location and the stopover range in between, obscures the immense physical challenge of what a migrant bird must overcome during its annual cycle.

First, Neotropical migrants, as with cranes and waterfowl, must quickly return to their nesting grounds from their southern wintering range to better compete with others for the best available nesting sites and food supplies. They must then obtain enough food not only for their own nutrition, egg production, and feeding of chicks but also for the post-breeding molt, all of which demand great reserves of energy. They build up these reserves by feeding, in competition with other birds, on energy-rich food such as seeds, fruit, and insect larvae.

When birds depart in the fall, a journey of hundreds or even a thousand miles or more often awaits them as they navigate by the solar compass, the stellar compass (for nocturnal migration), the earth's magnetic field, or visual landmarks such as rivers and coastlines. Birds will need to refuel at some point, and often their stopover arrival times have evolved to coincide with peak food availability. Songbirds typically will stop after a few hundred miles overland for several hours of rest and refuelling before continuing. Southbound migrants may stop along the northern Gulf Coast before heading out over the water. Long-distance migrants such as shorebirds can fly non-stop for up to 2,000 miles each way on their fall and spring journeys.

Uncertainty haunts migrating shorebirds and songbirds all along their travels. Strong headwinds make flight more difficult; spring cold fronts and thunderstorms over the gulf either kill birds outright or knock them off course, forcing them to land in unfamiliar territory. A well-used stopover location, rich in food one year, may be barren the next. The hawk in the tree, the bobcat in the bush, a drought-withered grassland, a dried-up prairie pothole, brightly lit glass-enclosed skyscrapers, broadcast towers, wind farms, or a marina that had been a salt marsh the year before, all stand between the migrant bird and its destination. Simple exhaustion may mean the end of the journey. According to some estimates, around half of the population of some species that depart their breeding grounds in the fall never return.

The importance of Bolivar Flats and High Island to migrating birds comes from not only their dense tapestries of food-rich habitat — oak woods, ponds, marshes, upland grasslands, mudflats, sandy beach — but also from their central location within the avenue of mass migration, on the northwestern coast of the Gulf of Mexico. Analysis of data from weather surveillance radar, a powerful tool for studying the speed, direction, altitude, and density of bird migration, clearly demonstrates this pivotal geography. Images on radar screens, scattered with a constellation of iridescent blips, eerily show vast flocks of birds moving through the atmosphere as they approach the coast.

Birds and radar share a history that goes back to the earliest days of the technology. Early in World War II both Allied and German radar operators, not realizing that birds could create radar echoes, would see sudden and unexpected images on the screen that they took to be enemy activity, such as boats and planes. These phantom signals were so convincing that fighter planes were scrambled, soldiers rushed to battle stations, and, on one occasion, an invasion alarm was sounded. Birds simply were not suspected as the cause. With improvements in radar technology and correlation with visual observation, the military eventually realized that birds in flight could be detected by radar and that such detections created serious interference with combat operations. Recognition of bird images on radar soon became a standard part of training for operators.

One of the early questions in ornithology concerned the route of migration of Neotropical migrants in relation to the Gulf of Mexico. Did birds fly over it — trans-gulf migration — or did they go overland around its eastern and western edges, in circum-gulf migration? Prior to the use of radar data in bird studies, the patterns of bird migration were largely inferred from the observations of amateur birdwatchers and professional biologists in the field. From the early 1900s, it was generally accepted that many Neotropical migrants flew north over the gulf in spring from the tropics. Much of the observational data, the timing and location of bird arrivals and departures along the coast, and the geographic distribution of species strongly suggested flights across the gulf.

A feisty debate erupted when George G. Williams, a professor at the then-Rice Institute, challenged the prevailing belief about trans-gulf migration, asserting that spring migrants came north only by an overland route from the Yucatan to either Mexico and Texas or to Cuba and Florida. He based this claim on an extensive analysis of bird records and his own observations, stating flatly that "[t]here is no direct evidence to show that birds migrating from regions south of us in spring actually cross the Gulf of Mexico." George H. Lowery Jr., of Louisiana State University, Williams's most prominent critic, called his claim a "fallacy" and his arguments "hopelessly misleading." The debate simmered as duelling articles flew back and forth in the academic journals. As it turned out, both sides were correct: Birds migrate by both the trans-gulf and circum-gulf routes in fall and spring.

Although Williams was an English professor and not a trained scientist, he was a respected amateur ornithologist who made important contributions to the knowledge of Texas bird life. While his claim of exclusive circum-gulf migration was disproven, he did leave a birding legacy: the outstanding collection of bird books that he spent years gathering for Rice's Fondren Library.

Modern radar studies have yielded huge advances in the understanding of bird migration. Much of the work on spring migration over the Gulf of Mexico has been conducted by the research team of Dr. Sidney A. Gauthreaux Jr., of the Radar Ornithology Laboratory at Clemson University, using a network of weather radar stations located along the coast. Their analysis reveals that birds that return north over the gulf do so across a broad front covering the coast from South Texas to the Florida Keys, with their departures originating predominantly from the Yucatan, the Caribbean, and Mexico, near the Bay of Campeche. Despite its many hazards, crossing the gulf assists migrants in a faster return to their breeding grounds. Birds leave at night and arrive on the Gulf Coast the following day, while an overland, circum-gulf route through Mexico and South Texas takes five to six days and exposes migrants to greater stress and predation.

The outstanding feature of trans-gulf spring migration is that, year after year, the greatest numbers of migrating birds arrive on a 500-mile stretch covering the upper Texas Coast and southwest Louisiana; the radar stations that consistently show the greatest density within this segment of the coast are Houston and Lake Charles. Migrants also come across the coast in Mississippi, Alabama, Florida, and South Texas, but in lower numbers.

Additional studies by Dr. Gauthreaux's team further suggest that the average longitude of peak arrival of migrants is near the 94th meridian, which runs between Sabine Pass and High Island. The sanctuaries of Bolivar Peninsula lie near the epicenter of the largest numbers of birds coming north across the gulf from the tropics.

The predominance of these areas as migrant destinations results to a large extent from the wind patterns that prevail over the gulf in spring. In the early part of the year, wind comes primarily from the east across the gulf, which would hinder the northbound migrant. Beginning around March, the onset of spring migration, winds come in from the south, steering birds toward the northern coast. If winds shift mildly to the east or west during migration, birds can compensate by altering their bearing and staying on their original course. But if strong east or west winds set in, the migrants that normally would go to the northern coast of the gulf land elsewhere. Ornithological records are full of observations of South Texas migrants making landfall in Florida and Florida migrants making landfall in South Texas.

Favorable wind conditions not only influence direction but also are extremely important to birds in flight by reducing the need for fuel consumption. Without the assistance of tailwinds, migrants are more likely to make landfall in a weakened condition since they must burn up extra fuel reserves to fight the elements. As a result, their chances for a prompt return to the breeding grounds and successful nesting are diminished. Even birds that depart the tropics in plump condition need favorable weather to reach land without becoming depleted of fat reserves. Wind aloft, rather than surface wind, is especially important since migrants fly at altitudes up to a mile and a half. If tailwinds are good, most birds will reach the coast in good condition and will continue inland, making their first stopover in coastal forests and grasslands.

The spring migration period of mid-March to late May each year (with the peak occurring in April) brings a rainbow of songbirds to High Island, especially during the event for which it is famous: a "fallout." This phenomenon occurs when birds encounter stormy weather as they approach the coast; strong headwinds and rain force them down by the thousands into the trees and onto the ground. If storms occur over the gulf, where there is no place to land, those birds not strong enough to fight the bad weather will not survive. (Fallouts have also been reported on offshore oil rigs and fishing boats.)

During a fallout the trees become vibrant with dozens of species of colorful songbirds that arrive after an overnight 500-mile, 18-hour journey across the gulf. Here a coastal habitat becomes a true refuge from the storm, a place to wait out the weather, to rest and refuel before continuing north. A branch of a large oak tree can contain a dozen or more species at one time. This stopover, whether in good weather or bad, hosts a brief and intimate gathering of birds that will soon scatter over the continent — the Rose-breasted Grosbeak to the Upper Midwest, the American Redstart to the Carolinas, the Prothonotary Warbler to East Texas, the Ovenbird to British Columbia. "They coincide for that brief time in the woods of High Island," says Ted Eubanks, "on their way to the forests of North America."

Although fall migration over the gulf has been studied much less than has spring migration, researchers have identified broad patterns of the southern journey to the tropics in relation to the Gulf of Mexico. As in spring migration, the routes of fall migration are greatly influenced by wind direction. With a strong cold front, coming from the north, birds will directly cross the gulf; an easterly wind may steer migrants slightly over the western gulf and more toward South Texas and northeastern Mexico; and without good tailwinds, migrants will tend to take the circum-gulf route. This is especially true of raptors, which need thermals for efficient long-distance flight and therefore seldom cross the gulf. Thermals don't develop over water as well as they do over a warm landscape.

Regardless of the migrant route taken, the Texas Coast witnesses a greater intensity of migrating birds in the fall than perhaps anywhere else in the country, as birds funnel down from the eastern forests, the American Great Plains, and Prairie Canada. Radar studies conducted by waterfowl researcher Dr. Bart Ballard of Texas A&M University–Kingsville show that fall migration passage rates — the numbers of migrants of all species within a given time span — along the Texas Coast are higher than those of any other area studied with radar in North America.

Many species of North American shorebirds and songbirds have severely declined in population, a trend that has been observed for four decades largely as a result of loss of nesting habitat. The data starkly reveal this disheartening trend. Both the Breeding Bird Survey (an international bird monitoring program administered by the U.S. Geological Survey and the Canadian Wildlife Service) and Partners in Flight (a consortium of numerous conservation organizations and governmental agencies) have shown that certain species of migratory shorebirds, such as the Red Knot and Wilson's Plover, have suffered declines of around 80 percent, with many other species also showing severe reductions in population. The U.S. Fish and Wildlife Service has identified other species, such as the Piping Plover and Long-billed Curlew, as "highly imperiled," with around two dozen other shorebirds listed as species of "high concern."

Songbirds have shown similar declines, with some species of forest and grassland birds plummeting 70 percent since 1970. This sad fading of our native songbirds manifests itself today at High Island, especially during fallouts. "A good day of birding at High Island today," says Gary Clark, nature columnist for the *Houston Chronicle* and author of the *Book of Texas Birds*, "would have been a poor day thirty years ago."

As described in the *United States Shorebird Conservation Plan* of the Manomet Center for Conservation Sciences, in Massachusetts, shorebirds present a special challenge to conservationists. These birds range widely over great distances that cross many countries; they rely on habitats, such as wetlands in the Great Plains and on shorelines, that are at risk from unpredictable weather and coastal development; they have low reproductive rates because of small egg clutches, high predation on the breeding grounds, and a tendency to not re-nest if the first nest fails; and, despite their journeys of thousands of miles, they often depend heavily on a small number of stopover locations. " Stopovers are very important to migrating birds," says Dr. Timothy Brush, a Neotropical migrant expert at the University of Texas–Rio Grande Valley, in Edinburg, Texas. "For example, the Wilson's Phalarope and Snowy Plover use the Salt Lakes of South Texas, while the Pectoral Sandpiper and White-rumped Sandpiper prefer inland freshwater ponds. The Western Sandpiper, the Willet, and many other species will use the tidal flats of the Laguna Madre. And a small but significant wintering population of the Red Knot occurs along the South Texas Coast."

The Red Knot provides what may be the most striking example of a bird's precarious dependence on a stopover. Each spring up to 80 percent of the Red Knot's population stops and feeds on horseshoe crab eggs along Delaware Bay before departing for its breeding grounds in Arctic polar deserts. Researchers have shown that food shortages that occur in some years from over-harvesting of crabs result in a lower survival rate and a population decline in the Red Knot. If horseshoe crabs were to disappear from Delaware Bay, the consequences to the Red Knot are all too obvious.

However, not all the bird survey data in recent years are grim. Many species of migrants have stabilized after sharp declines during the early 1980s and mid-1990s and some have shown healthy increases. But the trend of decline is broad enough across the Charadriiformes to demand continued vigilance by conservation groups to protect them. However, developing effective conservation strategies for shorebirds — or any animal species — requires detailed knowledge about their habitats and nesting requirements, migratory routes, nutritional needs, and vulnerabilities. Shorebirds, with their hemispheric reach, present a special problem: How does one study, monitor, and protect birds that nest and winter in remote locations that are 8,000 miles apart?

"There are large gaps in our understanding of shorebird populations," says Dr. Brad Andres, National Coordinator of the U.S. Fish and Wildlife Service's Shorebird Conservation Partnership. "It is very difficult to survey them in northern Canada." Researchers estimate that population trend data are totally lacking for around 25 percent of North American shorebirds. Among the birds that have been more thoroughly studied, "more species are declining than are increasing," Dr. Andres says.

If any one species best captures the peril and geographic sweep of the shorebirds, as well as an intimate dependence on the Texas Coast, it is the Hudsonian Godwit. This hefty black-and-chestnut bird is one of the least understood of the shorebirds, as scientists grasp its migratory odyssey only in the broadest terms. It has three separate breeding populations: Alaska, the northern Northwest Territories, and the shorelines of Hudson and James bays. In late summer, godwits begin to move south and stage in large numbers in south-central Canada. Then they embark on a non-stop 2,000-mile journey over the open Atlantic, the last of them reaching the Amazon Basin by September. They continue south into southern Brazil and northern Argentina, eventually gathering in large numbers at their final wintering locations in southern Chile and at land's end for the Americas, Tierra del Fuego.

In winter, godwits concentrate in only a few areas, such as Bahía San Sebastián, in Tierra del Fuego, and Isla Chiloé, in southern Chile. Here they feed and roost, before embarking, around February, on the long journey back to Alaska and northern Canada. Taking a different route in spring, Hudsonian Godwits head north to staging areas in Argentina and Brazil before undertaking what is believed to be a non-stop journey to northeastern Mexico and Texas. With most of the godwits migrating in late April and early May through the rice fields and flooded pastures of Galveston and Chambers counties, almost the entire global population of the Hudsonian Godwit passes through the Texas Coast. They then follow a direct path north over the Great Plains to their nesting ranges. After a few months, the godwits return to South America.

Although it nests and spends the winter at the opposite ends of the Western Hemisphere, the Hudsonian Godwit is not beyond the destructive grasp of human disturbance. Conservation plans developed in Canada and the US, which categorize the bird as "high concern," point to threats at all stages of the annual cycle. The godwit's small population — 77,000 by recent estimates — is compounded by the bird's heavy reliance on only a few nesting, staging, and wintering areas, many of which are under threat of development.

The Mackenzie River Delta, a prime nesting area, is the site of a proposed major oil and gas pipeline; Bahía San Sebastián, the location of the largest wintering flocks of the Hudsonian Godwit, has been proposed for a shipping and ferry terminal. Stopover pastures and wetlands from Texas to the Dakotas are always at the mercy of agriculture and the unpredictable Great Plains weather. The godwit's small population makes these threats to essential habitat even more disquieting. What happens to the Hudsonian Godwit may also be the fate of other shorebirds.

Bolivar Flats and High Island have recovered from their assaults by hurricanes Humberto and Ike. The gulf has healed the wounded shoreline with fresh sand and the shorebirds have been gathering once again. The trees on top of the salt dome still stand in dark-green glory, protected by the elevation of the land from the poison of surging salt water. Aerial photographs taken of High Island a few days after Hurricane Ike showed that the town had briefly become, quite literally, an island. Many trees were battered, but the warblers and orioles and all the others still return. As Houston Sliger, a long-time High Island resident told the author: "The birds don't care what the trees look like."

As important as the Bolivar sanctuaries are, they nevertheless belong to a larger coastal world that provides much food and shelter to migrating birds, perhaps more so than do the flats and High Island. The Laguna Madre and Baffin Bay host huge flocks of shorebirds each winter. The Columbia Bottomlands, a 250-square mile region of trees along the Brazos, Colorado, and San Bernard rivers, no doubt give spring refuge to more birds than does High Island.

The importance of Bolivar's sanctuaries to Texas and North American bird life, however, goes beyond providing material benefit to migrants. The mudflats and woodlands have played a role, in some ineffable way, in our understanding and appreciation of the life of birds. There was a time, not that many years ago, when the grandeur of our hemispheric migrants was not widely known. Bolivar Flats and High Island came about because of the efforts of concerned people who recognized their importance. They are accessible; they are visible; they have survived; they are simply *there*, not just for the birds but also for people who care about them; and they have been saved.

So now, brightly painted new houses again overlook the Gulf of Mexico. The ferries come and go. Highway 87 has been repaired. The communities have come alive with restaurants, hardware stores, storage units, shopping centers, a school, parks, fishing spots, bait camps, marinas, a post office, real estate offices, and more real estate offices. Life, both human and wild, has returned to the scoured landscape. Our shorebirds and songbirds — those sky pilgrims of two continents — once again have their haven in the sun.

Acknowledgments

TEXAS CAN CLAIM NOT only extraordinary bird life but also countless dedicated individuals devoted to its management and preservation. One of the greatest pleasures I enjoyed while working on this book was the opportunity to speak with many of them — private citizens, ranchers and landowners, university researchers, federal and state wildlife managers, conservationists, museum curators, librarians, archivists, authors, and many others. Without their generous sharing of not just their knowledge but also their experience and insight, I could not have written this book. But my first acknowledgement goes to my wife Middy Randerson, my biggest supporter and toughest critic. We met in the newsroom of *The Houston Post* when I was a nature columnist there and she was a writer in the Features department; she later became the Travel Editor. Her discerning eye and editorial insights led to a much better book; our spirited discussions over parts of the text generally concluded, correctly so, with my acknowledgment that her idea was better. My editors at Texas Tech University Press — Managing Director Joanna Conrad, Acquisitions Editor Travis Snyder, and Copy Editor Christie Perlmutter, as well as Senior Designer Hannah Gaskamp — all deserve special thanks for their outstanding guidance and editorial insight in making the book a reality.

Of the many others who assisted me I would like to give grateful recognition to the following: Gary Clark, The Woodlands, Texas, nature columnist for the *Houston Chronicle* and author of the *Book of*

Texas Birds, for his peer review of Chapter 1 and his early support and encouragement. Terry A. Rossignol, former Refuge Manager, Attwater Prairie Chicken National Wildlife Refuge, Eagle Lake, Texas, for peer review of Chapter 1, for access to refuge archives, and for taking me around the restricted areas of the refuge. Terry also greatly helped in understanding the many subtleties of the grouse's need for prairie grass. In addition, he facilitated a visit for me to the private landowner in Goliad who was participating in the captive-release program there. Ted L. Eubanks Jr., Fermata Inc. Austin, Texas, for his insights into Gulf coastal migration and the history of the establishment of the sanctuaries at Bolivar Flats and High Island. William P. Kuvlesky Jr., Texas A&M University–Kingsville, for arranging the rare opportunity of an overnight visit on the King Ranch to look for White-tailed Hawks; he is one of the few researchers to have conducted field studies on the ecology of this uncommon raptor and on the bird's relationship to land management practices. Dr. Kuvlesky also provided a peer-review of Chapter 2. Capt. Sally Black of the Baffin Bay Rod and Gun Club, Riviera, Texas, who took me out by boat on the Laguna Madre on a very cold December morning to see flocks of Redheads — only later to have the opportunity to explain what we were doing to Texas Park and Wildlife Department personnel on an airboat who stopped us to inquire about our activities. Warren Pulich, Texas State University, San Marcos, for peer review of Chapter 4 and for helping me to acquire at least a beginner's understanding of the complexities of seagrass ecology. Bart Ballard, Caesar Kleberg Wildlife Research Institute, Texas A&M University–Kingsville, for peer review of Chapter 4 and for his considerable insight into waterfowl migration. Tom Stehn, Aransas Pass, Texas, serious student of the blues and former longtime U.S. Fish and Wildlife Service Whooping Crane Coordinator at the Aransas National Wildlife Refuge, Austwell, Texas, for peer review of Chapter 3. Tom has a deep understanding of the ecology of *Grus americana* and of the special challenges this great bird faces for its survival, and we had memorable times together roaming around Houston's Third Ward looking for the ghost of Lightnin' Hopkins. Hannah Bailey, Director of the Attwater Prairie Chicken captive breeding program at the Houston Zoo, for showing me around their impressive facilities. Ronnie Howard, former hunting lease manager of San Tomas hunting camp, Encino Division, King Ranch, Falfurrias, Texas, for driving me around the King Ranch in search of the White-tailed Hawk and explaining the benefits of quail and cattle management to this raptor. Houston Sliger, longtime resident of High Island, Texas, who walked me through the town's history and took me on a personal tour of the oak woodland sanctuaries. Claud Kahla, longtime High Island resident, for his knowledge of the aftermath of Hurricane Ike and the history of the Bolivar sanctuaries. Shannon Tomkins, former outdoor writer for the *Houston Chronicle*, for being that rare outdoor journalist with a keen understanding of wildlife and the touch of a poet; Shannon provided valuable insight into the behavior of the Redhead. Stennie Meadours, San Leon, Texas, for her diligent and courageous efforts in the establishment of Bolivar Flats Shorebird Sanctuary. Winnie Burkett, Port Bolivar, Texas, for her hospitality at her beach home overlooking Bolivar Flats, her detailed knowledge of shorebird populations on the Texas Coast, and her recollections about Hurricane Ike. James B. Blackburn Jr., environmental attorney and one of my first professors at Rice University, who long ago taught that environmental conservation is not an obstacle to a healthy economy but its handmaiden. And special thanks to Madeline Camp, of Shoreline, Washington, who, as my research assistant while a student at Rice University, greatly assisted me in understanding rice agriculture and the wind farm program. Finally, to Herb Ward and Frank Fisher, Professors Emeriti in the Department of Ecology and Evolutionary Biology at Rice University, mentors and teachers, for their many years of friendship, inspiration, and wise counsel.

I would also like to extend my sincerest gratitude to the following individuals for their invaluable assistance:

Chapter 1, Attwater's Prairie Chicken: Clifton Carter, Soil Conservation Service (ret.), Victoria, Texas, personal anecdote about Valgene Lehmann. Ranch owner Morgan O'Connor and ranch foreman Stephan Schaar, O'Connor Ranch, Goliad County, Texas, Safe Harbors program in Goliad. Tim O'Connell, steward at the Texas City Prairie Preserve, Texas City, Texas. The staff at the Dolph Briscoe Center for American History at the University of Texas at Austin, Valgene Lehmann papers and references to early Texas explorers Frederic Law Olmstead, William Kennedy, and Mary Austin Holley. The staff at the Nesbitt Memorial Library, Columbus, Texas,

Thomas Waddell papers and 1927 Eagle Lake *Headlight* article. The staff at the Woodson Research Center at Fondren Library, Rice University, early newspaper articles and publications of Charles Emil Bendire on Attwater's Prairie Chicken. The staff at the Houston Public Library, Texas and Local History Collections, Julia Ideson Building, biographical information on Henry P. Attwater. Claudia Angle, bird curator at the National Museum of Natural History, Smithsonian Institution, Washington, DC, for showing me the original specimens of the prairie chicken collected by H. P. Attwater. The staff of Hollywood Cemetery, Houston, H.P. Attwater's burial records and grave site. Tim Anderson, U.S. Fish and Wildlife Service, Corpus Christi, private land programs for Attwater's Prairie Chicken. Jerry Gray, Agricultural Extension Agent, Refugio County, agriculture of Refugio County.

Chapter 2, White-tailed Hawk: Cliff Shackelford, ornithologist, Non-game and Rare Species Program, Texas Parks and Wildlife Department and Craig Farquhar, Endangered Species Grants Coordinator, Wildlife Division, Texas Parks and Wildlife Department, range and ecology. Dr. Farquhar also provided a peer review for Chapter 2. Dan Walker, biologist, Chaparral Wildlife Management Area, Cotulla, Texas, northern range of the hawk. Terry Blankenship, Director, Welder Wildlife Foundation, Sinton, Texas, ecology of South Texas avian specialties and peer review of Chapter 2. Kenneth Jacobson, Raptor Management Coordinator, Arizona Game and Fish Department, Phoenix, historic occurrence of the White-tailed Hawk in Arizona. Meg Streich, Conservation Education and Volunteer Program Coordinator, Welder Wildlife Foundation, Sinton, Texas, library facilities. Chuck Davis, Region 8 Director (Coastal Prairie), Texas Ornithological Society, breeding range. Lee Gaston, Wildlife Specialist and Assistant Manager, Brazoria National Wildlife Refuge, occurrence of the hawk on the refuge. Jimmy Laurent, Manager, Anahuac National Wildlife Refuge, occurrence of the hawk on the refuge. Kristin Madden, Bird Program Manager, Wildlife Management Division, New Mexico Department of Game and Fish, Santa Fe, NM, historic occurrence in New Mexico. Donna Dittmann, Secretary, Louisiana Ornithological Society Bird Records Committee, occurrence in Louisiana. Christopher Rustay, New Mexico Ornithological Society Bird Records Committee, occurrence in New Mexico. David Sarkozi, Secretary, Texas Ornithological Society, nesting range in Texas. Laramie Adams, Director of Public Affairs, Texas and Southwestern Cattle Raisers Association, Austin, challenges facing the modern rancher. David E. Brown, Arizona State University, Tempe, historic occurrence of the hawk in Arizona.

Chapter 3, Whooping Crane: Aaron Pearse, research wildlife biologist, U.S. Geological Survey Northern Prairie Wildlife Research Center, Jamestown, North Dakota, migration data. Lana Cortese, ecologist team lead, Parks Canada–Wood Buffalo National Park, Fort Smith, Northwest Territories, Canada, location of crane nesting areas. Mark Wayland, Environmental Stewardship Branch, Canadian Wildlife Service, Saskatoon, Saskatchewan, nesting ecology. Jim Blackburn, environmental attorney, Houston, Texas, freshwater inflow and legal issues of Aransas Project. Felipe Chavez-Ramirez, former director of conservation programs, Gulf Coast Bird Observatory, Lake Jackson, Texas, nesting, feeding and migration ecology. Greg Birkenfeld, former U.S. Fish and Wildlife Service, Refuge Manager, Aransas National Wildlife Refuge, Austwell, Texas, history of the refuge. Stuart Marcus, U.S. Fish and Wildlife Service, Refuge Manager, Trinity River National Wildlife Refuge, Liberty, Texas, access to archives of the Aransas National Wildlife Refuge. Wade Harrell, U.S. Fish and Wildlife Service, Whooping Crane Coordinator, Aransas National Wildlife Refuge, Austwell, Texas, crane wintering ecology and population distribution. Dr. Harrell also provided a peer review for Chapter 3. Capt. Tommy Moore, captain of tour boat The Skimmer, Fulton, Texas, for his always entertaining and informative trips to see the cranes. John Welder, Victoria, Texas, fifth-generation Texan, farmer, rancher, Welder Flats, land conservation and stewardship. Linda Spiro, Kelley Center for Government Information, Data, and Geospatial Service, Fondren Library, Rice University, Houston, Texas, copy of Executive Order 7784 establishing the Aransas Migratory Waterfowl Refuge. Norman Boyd, San Antonio Bay Ecosystem Leader, Coastal Fisheries, Texas Parks and Wildlife Department, Port O'Connor, Texas, ecology of blue crab. Daniel M. Alonso, Executive Director, San Antonio Bay Foundation, Seadrift, Texas, ecology of blue crab. Mark Fisher, Science Director, Coastal Fisheries Division, Texas Parks and Wildlife, Rockport, Texas, decline of blue crab. Barry K. Hartup, Director of Conservation Medicine, International Crane Foundation, Baraboo,

WI, infectious diseases and endoparasitism within the wild flock. Victor Emanuel, Victor Emanuel Nature Tours, Austin, Texas, development of nature tourism. Diane Probst, President/CEO, Rockport-Fulton Chamber of Commerce, economic impact of nature tourism. Diane and Al Johnson, Crane House, Lamar, Texas, crane tourism. Elizabeth Smith, Program Director, Texas Whooping Crane Project, International Crane Foundation, Lamar, Texas, introduced flocks. Selma Glasscock, Assistant Director, Rob and Bessie Welder Wildlife Foundation, Sinton, Texas, history of foundation and the Welder family will. Matt Rabbe, Senior Wildlife Biologist, U.S. Fish and Wildlife Service, Ecological Services Division, Wood River, Nebraska, Platte River migration habitat. Robert Love, Administrator, Coastal Non-game Resources Division, Louisiana Department of Wildlife and Fisheries, Baton Rouge, Louisiana crane re-introduction program. Stuart Macmillan, biologist, Wood Buffalo National Park, Fort Smith, Northwest Territories, Canada, hydroelectric and oil sands projects in Wood Buffalo National Park.

Chapter 4, Redhead: Kevin Kraai, Waterfowl Program Leader, Texas Parks and Wildlife Department, Canyon, Texas, duck population trends and conservation. Sally Black and Audrey Black, Baffin Bay Rod and Gun Club, Riviera, Texas, Redhead distribution on the Laguna Madre. J. Dale James, Manager of Conservation Planning, 13-state Southern Region, Ducks Unlimited, Jackson, MS, duck population trends. Kirby Brown, Conservation Outreach Biologist, Ducks Unlimited Texas, San Antonio, Texas, duck population trends. Dave Morrison, Small Game Program Director, Wildlife Division, Texas Parks and Wildlife Department, Austin, duck population trends and conservation. Todd Merendino, Manager, Conservation Program, Ducks Unlimited Texas, Richmond, Texas, waterfowl conservation. Jamie Feddersen, state waterfowl biologist, Florida Fish and Wildlife Conservation Commission, Tallahassee, FL, the Redhead in Florida. Neal D. Niemuth, Integrated Bird Conservation Scientist, Habitat and Population Evaluation Team (HAPET), U.S. Fish and Wildlife Service, Bismarck, North Dakota, Prairie Pothole Region. Thomas Michot, University of Louisiana at Lafayette, Redhead surveys along Gulf Coast. Jackie Robinson, Coastal Ecologist, and Russell Hooten, Wildlife Habitat Assessment Biologist, Texas Parks and Wildlife

Department, Corpus Christi, wind farm permitting issues. Kim Withers, Texas A&M University–Corpus Christi, seagrass ecology and plant succession. Larry Hindman, former Waterfowl Project Leader, Maryland Department of Natural Resources (ret.), Annapolis, the Redhead and Chesapeake Bay. Tim Fulbright, Caesar Kleberg Wildlife Research Institute, Texas A&M University–Kingsville, vegetation of the King Ranch. Paul Montagna, Texas A&M University–Corpus Christi, eolian environment, climate change effects on Texas Coast. Kyle Spiller, Texas Parks and Wildlife Department (ret.), naming of Laguna Madre. Faye Grubbs, Coastal Fisheries, Texas Parks and Wildlife Department, Corpus Christi, fisheries data from Laguna Madre. John W. (Wes) Tunnell Jr., Texas A&M University–Corpus Christi (ret.), history and ecology of the Laguna Madre. Kammie L. Kruse, U.S. Fish and Wildlife Service, Albuquerque, New Mexico, mid-winter survey data of Redheads along Gulf of Mexico.

Chapter 5, Migrant songbirds and shorebirds: Richard Gibbons, Conservation Director, Houston Audubon, Missouri City, Texas, peer review of Introduction and Chapter 5 and use of Bolivar Flats and High Island sanctuaries by migrant songbirds and shorebirds. Peggy Dillard, Special Collections Manager, Rosenberg Library, Galveston, Texas, history of Bolivar Flats Shorebird Sanctuary. John B. Anderson, Professor of Geology, Rice University, Houston, formation and evolution of barrier islands. Galveston County Judge Ray Holbrook (ret.), Santa Fe, Texas, establishment of Bolivar Flats Shorebird Sanctuary. Lisa Lepage, Western Hemisphere Shorebird Reserve Network, Manomet, MA, shorebird areas of hemispheric importance. Brad Andres, National Coordinator, U.S. Shorebird Conservation Partnership, U.S. Fish and Wildlife Service, Denver, CO, conservation issues of shorebirds. Timothy Brush, University of Texas–Rio Grande Valley, Edinburg, Texas, population trends and habitat needs of Neotropical migrants and peer review of Chapter 5; Robert Zink, Nebraska State Museum, Lincoln, Nebraska, evolution of avian migration. John Faaborg, University of Missouri–Columbia, origins of avian migration, conservation issues. Kimberly Vetter, Senior Communications Specialist, Texas Children's Hospital, editorial review of Introduction and Chapter 5.

And finally, a special word of recognition goes to another family member and faithful writing companion, my tabby cat Buster.

Works Consulted

INTRODUCTION

Anderson, J. B. *The Formation and Future of the Texas Coast.* College Station: Texas A&M University Press, 2007: xvi, 163.

Armstrong, N. E. "The Ecology of Open-bay Bottoms of Texas: A Community Profile." *U.S. Fish and Wildlife Service Biological Report* 85 (January 1987): 104.

Bates, R. L., and J. A. Jackson, eds. *Dictionary of Geological Terms.* Garden City, NY: Anchor Press/Doubleday, 1984: 571.

Britton, J. C., and B. Morton. *Shore Ecology of the Gulf of Mexico.* Austin: University of Texas Press, 1989: viii, 387.

California Birds Records Committee. "Official California Checklist." http://www.californiabirds.org/checklist.asp, 2019. Accessed May 14, 2019.

Carr, J. T., Jr. "The Climate and Physiography of Texas." *Report 53.* Austin: Texas Water Development Board, 1967: 27.

Chesser, R. T., K. J. Burns, C. Cicero, J. L. Dunn, A. W. Kratter, I. J. Lovette, P. C. Rasmussen, J. V. Remsen, Jr., D. F. Stotz, and K. Winker. 2019. Check-list of North American Birds (online). American Ornithological Society. http://checklist.aou.org/taxa

City of Houston, Planning and Development Department. "Historical Population: 1900–2013." https://www.houstontx.gov/planning/Demographics/docs_pdfs/Cy/coh_hist_pop.pdf, 2013. Accessed May

15, 2019.

Clark, G. *Book of Texas Birds.* College Station: Texas A&M University Press, 2016: x, 500.

Everitt, J. H., D. L. Drawe, and R. I. Leonard. *Trees, Shrubs & Cacti of South Texas.* Lubbock: Texas Tech University Press, 2002: xi, 249.

Hall, S. L., W. R. Wilder, and F. M. Fisher. "An Analysis of Shoreline Erosion along the Northern Coast of East Galveston Bay, Texas." *Journal of Coastal Research* 2(2) (1986): 173–79.

Houston Outdoor Nature Club, Ornithology Group. *A Birder's Checklist of the Upper Texas Coast*, 9th ed., May 2008.

Kloesel, K., B. Bartush, J. Banner, D. Brown, J. Lemery, X. Lin, C. Loeffler, G. McManus, E. Mullens, J. Nielsen-Gammon, M. Shafer, C. Sorensen, S. Sperry, D. Wildcat, and J. Ziolkowska. "Southern Great Plains." In *Impacts, Risks, and Adaptation in the United States: Fourth National Climate Assessment*, Volume II. Edited by D. R. Reidmiller, C. W. Avery, D. R. Easterling, K. E. Kunkel, K. L. M. Lewis, T. K. Maycock, and B. C. Stewart. U.S. Global Change Research Program, Washington, DC, 2018: 987–1035. doi: 10.7930/NCA4.2018.CH23

National Audubon Society. "Christmas Bird Count Statistics 2017–2018." https://www.audubon.org/news/the-118th-christmas-bird-count-summary, 2019. Accessed May 17, 2019.

National Oceanic and Atmospheric Administration. "National Coastal Population Report: Population Trends from 1970 to 2020." https://aamboceanservice.blob.core.windows.net/oceanservice-prod/facts/coastal-population-report.pdf, 2019. Accessed May 15, 2019.

Oberholser, H. C., and E. B. Kinkaid Jr., eds. *The Bird Life of Texas*, vol. 1. Austin: University of Texas Press, 1974: xxvii, 530. (See p. 114 for the citation for Norma C. Oates.)

Paine, J. G. "Subsidence of the Texas Coast: Inferences from Historical and Late Pleistocene Sea Levels." *Tectonophysics* 222 (1993): 445–58.

Paine, J. G., T. L. Caudle, and J. R. Andrews. *Shoreline Movement along the Texas Gulf Coast.* Austin: Bureau of Economic Geology, University of Texas at Austin. Report prepared for General Land Office under Contract No. 09-074-000, Work Order No. 7776, 2014: 52.

Peterson, R. T. *A Field Guide to the Birds of Texas and Adjacent States.* Boston: Houghton Mifflin Company, 1963: xxx, 304.

Texas Bird Records Committee. "Texas State List." http://www.texas-birdrecordscommittee.org/home/texas-state-list, 2019. Accessed May 14, 2019.

Texas Water Development Board. "Coastal Bend Population Projections." https://www.twdb.texas.gov/waterplanning/data/projections/2012/doc/Population/1StatePopulation.pdf, 2012. Accessed July 7, 2015.

Tompkins, Shannon. "Warmer Winters Trigger Changes in Texas Bays." *Houston Chronicle.* December 8, 2018. https://www.houston-chronicle.com/sports/outdoors/article/Warmer-winters-trigger-changes-in-Texas-bays-13451470.php. Accessed May 20, 2019.

United States Census Bureau. "Coastline Population Trends in the United States: 1960 to 2008." https://www.census.gov/prod/2010pubs/p25-1139.pdf, 2010. Accessed November 10, 2015.

United States Census Bureau. "New Census Bureau Population Estimates Show Dallas-Fort Worth-Arlington Has Largest Growth in the United States." https://www.census.gov/newsroom/press-releases/2018/popest-metro-county.html, 2018. Accessed May 15, 2019.

United States Census Bureau. "2018 State and National Population Estimates." https://www.census.gov/newsroom/press-kits/2018/pop-estimates-national-state.html, 2018. Accessed May 15, 2019.

CHAPTER 1

Aldrich, J. W. "Geographic Orientation of American Tetraonidae." *The Journal of Wildlife Management* 27(4) (1963): 528–45.

American Association for the Advancement of Science. "The Last Heath Hen." *Science* 32(4) (1931): 382–84.

Anderson, R. C. "Evolution and Origin of the Central Grassland of North America: Climate, Fire, and Mammalian Grazers." *Journal of the Torrey Botanical Society* 133(4) (2006): 626–47.

Applegate, R. D., C. K. Williams, and R. R. Manes. "Assuring the Future of Prairie Grouse: Dogmas, Demagogues, and Getting Outside the Box." *Wildlife Society Bulletin* 32(1) (2004): 104–11.

Attwater, H. P. "The Disappearance of Wild Life." *Bulletin of the Scien-*

tific Society of San Antonio 1(3) (1917): 47–60.

Attwater Prairie Chicken National Wildlife Refuge, Annual Narrative Reports, 1973–2014.

Bailey, V. *Biological Survey of Texas*. North American Fauna No. 25. Washington, DC: Government Printing Office, 1904: 222.

Bendire, C. E. "Description of a New Prairie Hen." *Forest and Stream* 50(20) (1893): 245.

———. "*Tympanuchus Americanus Attwateri Bendire*. Attwater's or Southern Prairie Hen." *The Auk* 11(2) (1894): 130–32.

Berg, W. E. "Epilogue: The Ghosts of Prairie Grouse Past." *Wildlife Society Bulletin* 32(1) (2004): 123–26.

Breckenridge, W. J. "The Booming of the Prairie Chicken." *The Auk* 46 (1929): 540–43.

Brennan, L. A., and W. P. Kuvlesky Jr. "North American Grassland Birds: An Unfolding Conservation Crisis?" *The Journal of Wildlife Management* 69(1) (2005): 1–13.

Chambers, B., and N. Childs. "Characteristics of U.S. Rice Farming." Economic Research Service, U.S. Department of Agriculture, Rice Situation and Outlook/RCS, 2000: 29–34.

Chamrad, A. D. "Effects of Fire and Grazing on Coastal Prairie Rangeland and Attwater's Prairie Chicken Habitat." PhD diss., College Station: Texas A&M University, 1971: xiii, 120.

Cogar, V. F. "Food Habits of Attwater's Prairie Chicken in Refugio County." PhD diss., College Station: Texas A&M University, 1980: iv, 119.

Department of the Interior. "Game Species Dying Out, But May Be Saved, Texan Says." P.N. 113083, September 5, 1940, 4.

Dethloff, H. C. "Rice Revolution in the Southwest: 1880–1910." *The Arkansas Historical Quarterly* 29(1) (1970): 66–75.

Diamond, D. D., and F. E. Smeins. "Remnant Grassland Vegetation and Ecological Affinities of the Upper Coastal Prairie of Texas." *The Southwestern Naturalist* 29(3) (1984): 321–34.

DiMare, M. I. "Effects of Lek Shape on Reproductive Behavior of Attwater's Prairie Chicken." PhD diss., College Station: Texas A&M University, 1991: x, 75.

Flickinger, E. L., and D. M. Swineford. "Environmental Contaminant Hazards to Attwater's Greater Prairie-Chicken." *The Journal of*

Wildlife Management 47(4) (1983): 1132–37.

Giese, M. W. "A Federal Foundation for Wildlife Conservation: The Evolution of the National Wildlife Refuge System, 1920–1968. PhD diss., American University, 2008: xiv, 519.

Hamerstrom, F., and F. Hamerstrom. "The Symposium in Review." *The Journal of Wildlife Management* 27(4) (1963): 869–87.

Handbook of Texas Online. "Henry Philemon Attwater." Austin: Texas State Historical Association. https://www.tshaonline.org/handbook/online. Accessed August 3, 2014.

Hannan, H. H., B. Callender, and J. D. Woodham. "Some Observations and Behavior Patterns of Prairie Chickens (*Tympanuchus cupido attwateri*) in Galveston County, Texas." *The Southwestern Naturalist* 9(4) (1964): 305–6.

Harcombe, P. A., G. N. Cameron, and E. G. Glumac. "Above-ground Net Primary Productivity in Adjacent Grassland and Woodland on the Coastal Prairie of Texas, USA." *Journal of Vegetation Science* 4 (1993): 521–30.

Henshaw, H. W. "Autobiographical Notes." *The Condor* 22 (1920): 3–10.

Horkel, J. D. "Cover and Space Requirements of Attwater's Prairie Chicken (*Tympanuchus cupido attwateri*) in Refugio County, Texas." PhD diss., College Station: Texas A&M University, 1979: xiii, 96.

Jacobs, B. F., J. D. Kingston, and L. F. Jacobs. *Annals of the Missouri Botanical Society* 86(2) (1999): 590–643.

Johnsgard, P. A. *Grouse and Quails of North America.* Lincoln: University of Nebraska Press, 1973: xx, 553.

Johnston, M. C. "Past and Present Grasslands of Southern Texas and Northeastern Mexico." *Ecology* 44(3) (1963): 456–66.

Jones, E. J. "Identification and Analysis of Lesser and Greater Prairie Chicken Habitat." *The Journal of Wildlife Management* 27(4) (1963): 757–78.

Jurries, R. W. *Attwater's Prairie Chicken.* F.A. Series No. 18. Austin: Texas Parks and Wildlife Department, 1979: 36.

Kirsch, L. M. "Habitat Management Considerations for Prairie Chicken." *Wildlife Society Bulletin* 2(3) (1974): 124–29.

Labuda, S. "The Present Status of Attwater's Prairie Chicken — 1989." Eagle Lake, TX: Attwater Prairie Chicken National Wildlife Refuge, 1989: 17.

Lee, J. "The Historical Geography of Rice Culture in the American South." PhD diss., Louisiana State University, 1988: xi, 225.

Lehmann, V. W. "Attwater's Prairie Chicken, Its Life History and Management." Washington, DC: U.S. Fish and Wildlife Society, *North American Fauna* 57 (1941): 63.

———. "Fire in the Range of Attwater's Prairie Chicken." In *Proceedings: Fourth Tall Timbers Fire Ecology Conference*, 127–43. Tallahassee: Tall Timbers Research Station, 1965: 279.

Lehmann, V. W., and R. G. Mauermann. "Status of Attwater's Prairie Chicken." *The Journal of Wildlife Management* 27(4) (1963): 713–25.

Lockwood, M. A., C. P. Griffin, M. E. Morrow, C. J. Randel, and N. J. Silvy. "Survival, Movements, and Reproduction of Released Captive-reared Attwater's Prairie-Chicken." *The Journal of Wildlife Management* 69(3) (2005): 1251–58.

Lockwood, M. A., M. E. Morrow, N. J. Silvy, and F. E. Smeins. "Spring Habitat Requirements of Captive-reared Attwater's Prairie Chicken." *Rangeland Ecology & Management* 58(3) (2005): 320–23.

Lutz, R. S., J. S. Lawrence, and N. J. Silvy. "Nesting Ecology of Attwater's Prairie Chicken." *The Journal of Wildlife Management* 58(2) (1994): 230–33.

Mayr, E. "History of the North American Bird Fauna." *The Wilson Bulletin* 58(1) (1946): 3–41.

Merriam, C. H. "Charles E. Bendire." *Science* 5(111) (1897): 261–62.

Morrow, M. E. "Ecology of Attwater's Prairie Chicken in Relation to Land Management Practices on the Attwater Prairie Chicken National Wildlife Refuge." PhD diss., Texas A&M University, 1986: x, 100.

———. "Evidence Continues to Mount Linking Red Imported Fire Ant to Attwater's Struggles." Eagle Lake, TX: Attwater Prairie Chicken National Wildlife Refuge. *The Boomer* 2(2) (2013): 3–4.

Morrow, M. E., R. S. Adamcik, J. D. Friday, and L. B. McKinney. "Factors Affecting Attwater's Prairie-Chicken Decline on the Attwater Prairie Chicken National Wildlife Refuge." *Wildlife Society Bulletin* 24(4) (1996): 593–601.

Nature Conservancy of Texas. "Update of the Texas City Prairie Preserve Conservation Area Plan." Supplement to the 1999 Conservation Plan, 2003: ii, 27.

Oberholser, H. C. "Obituary of Henry Philemon Attwater." *The Auk* 49 (1932): 144–45.

Oberholser, H. C., and E. B. Kinkaid Jr., eds. *The Bird Life of Texas*, vol. 1. Austin: University of Texas Press, 1974: xxvii, 530.

Peterson, M. J., and N. J. Silvy. "Reproductive Stages Limiting Productivity of the Endangered Attwater's Prairie Chicken." *Conservation Biology* 10(4) (1996): 1264–76.

———. "Spring Precipitation and Fluctuation in Attwater's Prairie-Chicken Numbers: Hypotheses Revisited." *The Journal of Wildlife Management* 58(2) (1994): 222–39.

Phillips, E. H. "The Gulf Coast Rice Industry." *Agricultural History* 25(2) (1952): 91–96.

Pierce, F. J. "The Prairie Chicken in East Central Iowa." *The Wilson Bulletin* 34(2) (1922): 100–106.

Pratt, A. C. "Evaluation of the Reintroduction of Attwater's Prairie Chicken in Goliad County, TX." Master's thesis, Texas A&M University–Kingsville, 2010: xiv, 75.

The Rice Journal, vol. 4, no. 6. (1901). Quotation obtained from Lee, J. 1988. "The Historical Geography of Rice Culture in the American South." PhD diss., Louisiana State University, 1988: xi, 225.

Robbins, M. B., A. T. Peterson, and M. A. Ortega-Huerta. "Major Negative Impacts of Early Intensive Cattle Stocking on Tallgrass Prairies: The Case of the Greater Prairie-Chicken (*Tympanuchus cupido*)." *North American Birds* 56(2) (2002): 239–44.

Robel, R. J. "Significance of Booming Grounds of Greater Prairie Chicken." *Proceedings of the American Philosophical Society* 111(2) (1967): 109–14.

———. "Summary Remarks and Personal Observations by an Old Hunter and Researcher." *Wildlife Society Bulletin* 32(1) (2004): 119–22.

Robel, R. J., and W. B. Ballard Jr. "Lek Social Organization and Reproductive Success in the Greater Prairie Chicken." *American Zoologist* 14 (1974): 121–28.

Samson, F. B., F. L. Knopf, and W. R. Ostlie. "Great Plains Ecosystems: Past, Present, and Future." *Wildlife Society Bulletin* 32(1) (2004): 6–15.

Silvy, N. J., and C. A. Hagen. "Introduction: Management of Imperiled Prairie Grouse Species and Their Habitat." *Wildlife Society Bulletin* 32(1) (2004): 2–5.

Silvy, N. J., M. J. Peterson, and R. R. Lopez. "The Cause of the Decline of the Pinnated Grouse: The Texas Example." *Wildlife Society Bulletin* 32(1) (2004): 16–21.

Snyder, J. W., E. C. Pelren, and J. A. Crawford. "Translocation Histories of Prairie Grouse in the United States." *Wildlife Society Bulletin* 27(2) (1999): 428–32.

Storch, Ilse. *Grouse: Status Survey and Conservation Action Plan 2000–2004.* WPA/BirdLife/SSC Grouse Specialist Group, 2000.

Tewes, C. "Valgene Lehmann — Early Pioneer in Wildlife Management." Caesar Kleberg Wildlife Research Institute, Texas A&M University–Kingsville. Special Publication No. 3, 2014: 10.

Texas A&M University Extension System. "Texas Rice." Vol. 5, no. 2 (2005): 12.

Texas Legislature: House of Representatives. *Journal of the House of Representatives of the First Called Session of the Forty-fifth Legislature.* May 27, 1937, 194. https://texashistory.unt.edu/ark:/67531/metapth145965/m1/501/?q=prairie chicken. Accessed June 7, 2014.

Texas Legislature: Senate. *Journal of the Senate, State of Texas, Regular Session, Thirty-Ninth Legislature, Legislative Document.* March 19, 1925, 687. http://texashistory.unt.edu/ark:/67531/metapth142172/. Accessed June 7, 2014.

U.S. Fish and Wildlife Service. "Attwater's Prairie Chicken Recovery Plan." Albuquerque, NM: Region 2, U.S. Fish and Wildlife Service, 1993: vi, 48.

U.S. Fish and Wildlife Service. "Management Guidelines for Attwater's Prairie Chicken." https://tpwd.texas.gov/publications/pwdpubs/media/pwd_bk_w7000_0013_attwaters_prairie_chicken_mgmt.pdf, n.d. Accessed May 7, 2014.

Wells, P. V. "Postglacial Vegetational History of the Great Plains." *Science* 167(3925) (1970): 1574–82.

CHAPTER 2

Actkinson, M. A. "Productivity and Nest-Site Selection of a Breeding Raptor Community in South Texas." Master's thesis, Texas A&M

University–Kingsville, 2006: x, 85.

Actkinson, M. A., W. P. Kuvlesky Jr., C. W. Boal, L. A. Brennan, and F. Hernandez. "Nesting Habitat Relationships of Sympatric Crested Caracaras, Red-Tailed Hawks, and White-Tailed Hawks in South Texas." *Wilson Journal of Ornithology* 119(4) (2007): 570–78.

Allen, J. A. "In Memoriam: George B. Sennett." *The Auk* 18 (1901): 11–23.

Asher, M. G. "Effects of Soil Amendments on Native and Exotic Grasses of the South Texas Coastal Plain." Master's thesis, Texas A&M University–Kingsville, 2003: 93.

Audubon Christmas Count Data. "Current Year Results by Species, 2014." Accessed online January 28, 2015.

Bailey, F. M. "Meeting Spring Half Way." *The Condor* 18 (1916): 183–90.

Baker, D. L. and F. S. Guthery. "Effects of Continuous Grazing on Habitat and Density of Ground-Foraging Birds in South Texas." *Journal of Range Management* 43(1) (1990): 2–5.

Bechard, M. J. "Effect of Vegetative Cover on Foraging Site Selection by Swainson's Hawk." *Condor* 84 (1982): 153–59.

Bildstein, K. L. "Raptor Migration in the Tropics: Patterns, Processes, and Consequences." *Ornitologia Neotropical* 15 (Suppl. (2004)): 83–99.

Bildstein, K. L., J. Zalles, J. Ottinger, and K. McCarty. "Conservation Biology of the World's Migratory Raptors: Status and Strategies." In *Raptors at Risk*. Edited by R. D. Chancellor and B.-U. Meyburg. World Working Group on Birds of Prey, Hancock House, 2000: 573–90.

Breninger, G. F. "White-Tailed Hawk in Arizona." *The Auk* 16 (1899): 352.

Brennan, L. A., S. DeMaso, F. Guthery, J. Hardin, C, Kowaleski, S. Lerich, R. Perez, M. Porter, D. Rollins, M. Sams, T. Trail and D. Wilhelm. "Where Have All the Quail Gone? A Proactive Approach to Restoring Quail Populations by Improving Wildlife Habitat." In *The Texas Quail Conservation Initiative*. Edited by R. Perez, T. Trail, D. Rollins, M. McMurray, and R. Macdonald. Texas Quail Council. Austin: Texas Parks and Wildlife Department, 2005: 22.

Brennan, L. A., and W. P. Kuvlesky Jr. "North American Grassland Birds: An Unfolding Conservation Crisis?" *Journal of Wildlife Man-*

agement, 69(1) (2005): 1–13.

Britton, C. M., S. Rideout-Hanzak, and S. D. Brown. "Effects of Burns Conducted in Summer and Winter on Vegetation of Matagorda Island, Texas." *The Southwestern Naturalist* 55(2) (2010): 193–202.

Brooks, A. "Some Notes on the Birds of Brownsville, Texas." *The Auk* 50 (1933): 59–63.

Brown, D. E., and R. L. Glinski. "The Status of the White-tailed Hawk in Arizona and Sonora, Mexico." *Journal of the Arizona-Nevada Academy of Science* 41(1) (2009): 8–15.

Brush, T. *Nesting Birds of a Tropical Frontier: The Lower Rio Grande Valley of Texas*. College Station: Texas A&M University Press, 2005: xiv, 245.

Box, T. W., and F. W. Gould. "An Analysis of the Grass Vegetation of Texas." *The Southwestern Naturalist* 3(1) (1958): 124–29.

Campbell-Kissock, L., L. H. Blankenship, and L. D. White. "Grazing Management Impacts on Quail during Drought in the Northern Rio Grande Plain." *Journal of Range Management* 37(5) (1984): 442–46.

Carroll, J. L. "Notes on the Birds of Refugio County, Texas." *The Auk* 17(4) (1900): 337–48.

Carter, P. S., D. Rollins, and C. B. Scott. "Initial Effects of Prescribed Burning on Survival and Nesting Success of Northern Bobwhite in West-Central Texas." In *Quail V: Proceedings of the Fifth National Quail Symposium*. Edited by S. J. DeMaso, W. P. Kuvlesky Jr., F. Hernández, and M. E. Berger. Austin: Texas Parks and Wildlife Department, 2002: 129–34.

Chavez-Ramirez, F. and F. G. Prieto. "Effects of Prescribed Fires on Habitat Use by Wintering Raptors on a Texas Barrier Island Grassland." *Journal of Raptor Research* 28(4) (1994): 262–65.

Cottam, C., and P. Knappen. "Food of Some Uncommon Birds." *The Auk* 56 (1939): 138–69.

Darwin, F., ed. *The Autobiography of Charles Darwin and Selected Letters*. New York: Dover Publications, Inc., 1958: 365.

Deane, R. "Extracts from the Field Notes of George B. Sennett." *The Auk* 404 (1923): 626–33.

Dittmann, D. L., S. W. Cardiff, and R. DeMay. *Louisiana Raptors: Birds of Prey*. Thibodaux, LA: Barataria-Terrebonne National Estuary

Program, n.d., 53.

Ditto, L. R. "Observations on Nesting White-tailed Hawks." *Raptor Research* 17(3) (1983): 91.

Drawe, D. L., and T. W. Box. "Forage Ratings for Deer and Cattle on the Welder Wildlife Refuge." *Journal of Range Management* 21(4) (1968): 225–28.

Farquhar, C. C. "Ecology and Breeding Behavior of the White-tailed Hawk on the Northern Coastal Prairies of Texas." PhD diss., Texas A&M University, 1968: x, 71.

——. "Individual and Intersexual Variation in Alarm Calls of the White-tailed Hawk." *The Condor* 95(1) (1993): 234–39.

——. "White-tailed Hawk." In *The Birds of North America*, No. 30. Edited by A. Poole, P. Stettenheim, and F. Gill. Philadelphia: The Academy of Natural Sciences. Washington, DC: The American Ornithologists' Union, 1992: 20.

Flanders, A. A., W. P. Kuvlesky Jr., D. C. Ruthven, III, R. E. Zaiglin, R. L. Bingham, T. E. Fulbright, F. Hernández, and L. A. Brennan. "Effects of Invasive Exotic Grasses on South Texas Rangeland Breeding Birds." *The Auk* 123(1) (2006): 171–82.

Forgason, C. A., and T. E. Fulbright. "Cattle, Wildlife, and Range Management on King Ranch over the Years." *Ranch Management: Integrating Cattle, Wildlife, and Range.* Kingsville, TX: King Ranch, 2003: 9–21.

Friedmann, H. "Notes on the Birds Observed in the Lower Rio Grande Valley of Texas during May, 1924." *The Auk* 42 (1925): 537–54.

Fulbright, T. E., and F. C. Bryant. *The Last Great Habitat.* Special Publication No. 1. Caesar Kleberg Wildlife Research Institute, Texas A&M University–Kingsville, 2004: 33.

Fulbright, T. E., and F. C. Bryant. "The Wild Horse Desert: Climate and Ecology." In *Ranch Management: Integrating Cattle, Wildlife, and Range.* Edited by C. A. Forgason, F. C. Bryant, and P. C. Genho. King Ranch Institute: King Ranch Symposium, 2003: 35–58.

Fulbright, T. E., D. D. Diamond, J. Rappole, and J. Norwine. "Coastal Sand Plain of Southern Texas." *Rangelands* 12(6) (1990): 337–40.

Graham, D. *The Kings of Texas: The 150-Year Saga of an American Ranching Empire.* Hoboken, NJ: John Wiley and Sons, 2003: xii, 289.

Griscom, L., and M. S. Crosby. "Birds of the Brownsville Region, Southern Texas." *The Auk* 42(3) (1925): 432–40.

Guthery, F. S. "A Philosophy of Habitat Management for Northern Bobwhites." *The Journal of Wildlife Management* 61(2) (1997): 291–301.

Haines, F. *The Buffalo: The Story of American Bison and Their Hunters from Prehistoric Times to the Present.* Norman: University of Oklahoma Press, 1995: ix, 244.

Hanselka, C. W. "Buffelgrass: South Texas Wonder Grass." *Rangelands* 10(6) (1988): 279–81.

———. ed. *Prescribed Range Burning in the Coastal Prairie and the Eastern Rio Grande Plains of Texas.* College Station: Texas Agricultural Extension Service, 1980: 128.

Hanselka, C. W., and F. S. Guthery. *Bobwhite Quail Management in South Texas.* College Station: Texas Agricultural Extension Service, 1991: 8.

Haralson, C. L. "Breeding Ecology Nest Site Selection and Human Influence of White-tailed Hawks on the Texas Barrier Islands." Master's thesis, Texas Tech University, 2008: xii, 103.

Haucke, H. H. "Predation by a White-tailed Hawk and a Harris Hawk on a Wild Turkey Poult." *The Condor* 73(4) (1971): 475.

Heredia, B., and W. S. Clark. "Kleptoparasitism by White-tailed Hawk (*Buteo albicaudatus*) on Black-Shouldered Kite (*Elanus caeruleus leucurus*) in Southern Texas." *Journal of Raptor Research* 18 (1984): 30–31.

Hernández, F., F. S. Guthery, and W. P. Kuvlesky Jr. "The Legacy of Bobwhite Research in South Texas." *The Journal of Wildlife Management* 66(1) (2002): 1–18.

Holechek, J. L., R. Valdez, S. D. Schemnitz, R. D. Pieper, and C. A. Davis. "Manipulation of Grazing to Improve or Maintain Wildlife Habitat." *The Wildlife Society Bulletin* 10(3) (1982): 204–10.

Jahrsdoefer, S. E., and D. M. Leslie Jr. "Tamaulipan Brushland of the Lower Rio Grande Valley of South Texas: Description, Human Impacts, and Management Options." U.S. Fish and Wildlife Service. *Biological Report* 88(36) (1988): 63.

Jensen, W. J. "The Abundance and Distribution of Falconiforms in the Western and Central Llanos of Venezuela." Master's thesis,

University of New York, 2003: v, 57.

Johnston, M. C. "Past and Present Grasslands of Southern Texas and Northeastern Mexico." *Ecology* 44(3) (1963): 456–66.

Kane, D. F. "Resource Partitioning and Nesting Ecology by a South Texas Raptor Assemblage." PhD diss., Texas A&M University–Kingsville, 2012: xv, 168.

Kelton, E. "Ranching in a Changing Land." *The Texas Almanac.* https://texasalmanac.com/, 2006. Accessed December 26, 2014.

Kuvlesky, W. P., Jr., T. E. Fulbright, and R. Engel-Wilson. "The Impact of Invasive Exotic Grasses on Quail in the Southwestern United States." In *Quail V: The Fifth National Quail Symposium.* Edited by S. J. DeMaso, W. P. Kuvlesky Jr., F. Hernández, and M. E. Berger. Austin: Texas Parks and Wildlife Department, 2002: 118–28.

Langschied, T. M. "Temporal Variation in Avian Communities in Southern Texas." Master's thesis, Texas A&M University–Kingsville, 1994: xii, 120.

Lasley, G., C. Sexton, M. Lockwood, C. Shackelford, and W. Sekula. "Texas Region." Texas Ornithological Society. *Field Notes* 52(1) (1998): 86–92.

Lea, T. *The King Ranch.* Two vols. New York: Little Brown and Company, 1957: 838.

LeFranc, M. N., Jr., and Brian A. Millsap. "A Summary of State and Federal Agency Raptor Management." *Wildlife Society Bulletin* 12(3) (1984): 274–82.

Lehmann, V. W. "Bobwhite Quail Reproduction in Southwestern Texas." *The Journal of Wildlife Management* 10(2) (1946): 111–23.

———. "Mobility of Bobwhite Quail in Southwestern Texas." *The Journal of Wildlife Management* 10(2) (1946): 124–36.

Lehmann, V. W., and H. Ward. "Some Plants Valuable to Quail in Southwestern Texas." *The Journal of Wildlife Management* 5(2) (1941): 131–35.

Lockwood, M. W., and B. Freeman. *The TOS Handbook of Texas Birds.* College Station: Texas A&M University Press, 2004: xxviii, 261.

Lusk, J. L., F. S. Guthery, R. R. George, M. J. Peterson, and S. J. DeMaso. "Relative Abundance of Bobwhites in Relation to Weather and Land Use." *The Journal of Wildlife Management* 66(4) (2002): 1040–51.

Lyons, R. K., and B. Rector, B. *Mesquite Ecology and Management*. College Station: Texas A&M AgriLife Extension Service, 2009: 7.

Mader, W. J. "Notes on Nesting Raptors in the Llanos of Venezuela." *The Condor* 83(1) (1981): 48–51.

Martinez-Gomez, J. E. "Raptor Conservation in Vera Cruz, Mexico." *Journal of Raptor Research* 26(3) (1992): 184–88.

McMahan, C. A., R. G. Frye, and K. L. Brown. *The Vegetation Types of Texas Including Cropland: An Illustrated Synopsis*. Austin: Texas Parks and Wildlife Department, 1984: 40.

Merrill, J. C. *Notes on the Ornithology of Southern Texas, being a List of Birds Observed in the Vicinity of Fort Brown, Texas from February, 1876 to June, 1878*. Washington, DC: Government Printing Office, Proceedings of the United States National Museum, 1878: 118–73.

Milgalter, R. A. "The Breeding Ecology, Nesting Habitat Preference, and Territory Occupancy of White-tailed Hawks, Red-Tailed Hawks, and Crested Caracaras in South Texas." Master's thesis, Texas A&M University–Kingsville, 2011: xii, 147.

Mix, K. D. "The Effects of Summer Prescribed Fire on Vegetation, Invertebrate and Avian Communities." Master's thesis, Texas A&M University–Kingsville, 2004: vii, 84.

Monday, J. C., and B. B. Colley. *Voices from the Wild Horse Desert: The Vaquero Families of the King and Kenedy Ranches*. Austin: University of Texas Press, 1997: xxxvii, 265.

Monteiro Granzinolli, M. A., and J. C. Motta-Junior. "Small Mammal Selection by the White-tailed Hawk in Southeastern Brazil." *Wilson Journal of Ornithology* 118(1) (2006): 91–98.

Morrison, M. L. "Breeding Characteristics, Eggshell Thinning, and Population Trends of White-tailed Hawks in Texas." *Bulletin of the Texas Ornithological Society* 11(2) (1978): 35–40.

Motta-Junior, J. C. "Consumption of a Ringed Kingfisher (*Megaceryle torquata*) by a White-tailed Hawk (*Buteo albicaudatus*) in Southeastern Brazil." *Journal of Raptor Research* 38(2) (2004): 191.

National Audubon Society. "Christmas Count Data, White Tailed-Hawk." http://netapp.audubon.org/CBCObservation/Reports/HistoricalResultsBySpecies.aspx?rf=EXCEL&sy=118&ey=118&sr=US-TX&sp=794&sc=True, 2017. Accessed August 23, 2019.

Neff, J. A. "Notes on Some Birds of Sonora, Mexico." *The Condor* 49(1)

(1947): 32–34.

New Mexico Ornithological Society. "Florence Merriam Bailey, Ornithologist." www.nmbirds.org/wp-content/florence-m-bailey-1.pdf. Accessed February 9, 2015.

Niven, D. L., and G. S. Butcher. "Status and Trends of Wintering Coastal Species along the Northern Gulf of Mexico, 1965–2011." *American Birds* 65 (2011): 12–19.

Oberholser, H. C., and E. B. Kinkaid Jr., eds. *The Bird Life of Texas*, vol. 1. Austin: University of Texas Press, 1974: xxvii, 530.

Olivo, C. "Fall Migration of the White-tailed Hawk in Central Bolivia. *Journal of Raptor Research* 37(1) (2003): 63–64

Ortega-S., J. A., and F. C. Bryant. "Cattle Management to Enhance Wildlife Habitat in South Texas." *Wildlife Management Bulletin* No. 6. Caesar Kleberg Wildlife Research Institute, Texas A&M University–Kingsville, 2005: 11.

Palmer, W. E., T. M. Terhune, T. Dailey, D. McKenzie, and J. Doty. *Executive Report.* National Bobwhite Conservation Initiative 2.0, 2011: 22.

Presley, A. L. "Antiquity and Paleoenvironment of the Tamaulipan Biotic Province of Southern Texas: The Zooarchaeological Perspective." Master's thesis, Texas A&M University, 2003: 161.

Preston, C. R. "Distribution of Raptor Foraging in Relation to Prey Biomass and Habitat Structure." *The Condor* 92 (1990): 107–12.

Purvis, J. *Small Game Harvest Survey Results 1994–1995 thru [sic] 2013–2014.* Texas Parks and Wildlife Department, 2014: 39.

Reynolds, M. C. "Effects of Burning on Birds in Mesquite-Grasslands." Master's thesis, University of Arizona, 1997: 71.

Roads, K. "Why Birds Are So Named." *The Wilson Bulletin* 24(1) (1912): 27–33.

Ruthven, D. C. III, T. E. Fulbright, S. L. Beasom, and Eric C. Hellgren. "Long-term Effects of Root Plowing on Vegetation in the Eastern South Texas Plains." *Journal of Range Management* 46(4) (1993): 351–54.

Ruvalcaba-Ortega, I., and J. I. González-Rojas. "New Records for Coahuila from a Riparian Bird Community in Northern Mexico." *The Southwestern Naturalist* 54(4) (2009): 501–9.

Sennett, G. B. "Some Undescribed Plumages of North American

birds." *The Auk* 4(1) (1887): 24–28.

Smith, A. P. "Miscellaneous Bird Notes from the Lower Río Grande." *The Condor* 12 (1910): 93–103.

Smith, F. S. "Texas Today: A Sea of the Wrong Grasses." *Ecological Restoration* 28(2) (2010): 112–17.

Spears, G. S., F. S. Guthery, S. M. Rice, S. J. Demaso, and B. Zaiglin. "Optimum Seral Stage for Northern Bobwhites as Influenced by Site Productivity." *The Journal of Wildlife Management* 57(4) (1993): 805–11.

Stevenson, J. O., and L. H. Meitzen. "Behavior and Food Habits of Sennett's White-tailed Hawk in Texas." *The Wilson Bulletin* 58(4) (1946): 198–205.

Tewes, M. E. "Opportunistic Feeding by White-tailed Hawks at Prescribed Burns." *The Wilson Bulletin* 96(1) (1984): 135–36.

Texas Parks and Wildlife. "Rare, Threatened, and Endangered Species of Texas." https://tpwd.texas.gov/gis/rtest/, 2019. Accessed May 17, 2019.

Texas State Historical Association. "Ranching," "Barbed Wire," "South Texas Plains," and "Rio Grande Valley." *Handbook of Texas Online.* https://www.tshaonline.org/handbook/online. Accessed September 20, 2014.

Tinkler, D. E. "Historical Change in Crop Distributions in Texas and its Potential Application for Modeling Wildlife Distribution." PhD diss., Texas Tech University, 2004: xiii, 90.

Turner, J. W. "Effects of Raptor Abundance on Northern Bobwhite Survival and Habitat Use in Southern Texas." Master's thesis, Texas A&M University–Kingsville, 2008: xiii, 53.

Tweit, R. C. "White-tailed Hawk." In *The Texas Breeding Bird Atlas.* College Station and Corpus Christi: Texas A&M University System. https://txtbba.tamu.edu/, 2008. Accessed October 6, 2014.

Vickery, P. D., and J. R. Herkert, eds. "Ecology and Conservation of Grassland Birds of the Western Hemisphere." *Studies in Avian Biology* no. 19. J. T. Cooper Ornithological Society, 1995: 299.

Villaseñor, J. F. "Habitat Use and the Effects of Disturbance on Wintering Birds Using Riparian Habitats in Sonora, Mexico." PhD diss., University of Montana–Missoula, 2006.

Voous, K. H. "Distribution and Variation of the White-tailed Hawk

(*Buteo albicaudatus*)." *Beaufortia* 15(208) (1968): 195–208.

White-tailed Hawk (*Geranoaetus albicaudatus*). In *Neotropical Birds Online*. Edited by T. S. Schulenberg. Cornell Lab of Ornithology, Ithaca, NY. Retrieved from Neotropical Birds Online: https://neotropical.birds.cornell.edu/Species-Account/nb/species/whthaw

Williams, C. K., F. S. Guthery, R. D. Appelgate, and M. J. Peterson. "The Northern Bobwhite Decline: Scaling Our Management for the Twenty-first Century." *Wildlife Society Bulletin* 32(3) (2004): 861–69.

CHAPTER 3

Allen, R. P. "Additional Data on the Food of the Whooping Crane." *The Auk* 71(2) (1954): 198.

———, ed. *A Report on the Whooping Crane's Northern Breeding Grounds.* New York: National Audubon Society, 1956.

———. *The Whooping Crane.* National Audubon Society Research Report no. 3. New York: National Audubon Society, 1952.

Aransas National Wildlife Refuge annual narrative reports. http://catalog.data.gov/dataset?q=Aransas+National+Wildlife+Refuge&sort=score+desc%2C+name+asc. Dat.Gov website accessed June 10, 2015.

Aransas National Wildlife Refuge, Whooping Crane Survey Results: Winter 2014–2015 survey data. https://www.fws.gov/uploaded-Files/WHCR%20Update%20Winter%202014-2015.pdf. Accessed November 25, 2019.

Aransas National Wildlife Refuge, Whooping Crane Survey Results: Winter 2018–2019. file:///C:/Users/brobison/Downloads/WHCR_Winter_Update_2018_2019%20(1).pdf. Accessed November 25, 2019.

Armbruster, M. J. "Characterization of Habitat Used by Whooping Cranes during Migration." U.S. Fish and Wildlife Service. *Biological Report* 90(4) (1990): 16.

Austin, J. E., and A. L. Richert. *A Comprehensive Review of Observational and Site Evaluation Data of Migrant Whooping Cranes in the United States, 1943–1999.* Final report to U.S. Fish and Wildlife Service, Region 6, and Nebraska Game and Parks Commission. Jamestown, ND. U.S. Geological Survey, Northern Prairie Wildlife

Research Center, 2001.

———. "Patterns of Habitat Use by Whooping Cranes during Migration: Summary from 1877–1999 Evaluation Data." *Proceedings of the North American Crane Workshop* 9 (2005): 79–104.

Behl, M. "March of the Mangroves: Studying the Effects of Mangrove Expansion on Coastal Wildlife." *Texas Shores* 42(1) (2014): 25–27.

Bergeson, D., B. W. Johns, and G. L. Holroyd. "Mortality of Whooping Crane Colts in Wood Buffalo National Park, Canada, 1997–1999." *Proceedings North American Crane Workshop* 8 (2000): 6–10.

Binkley, C. S., and R. S. Miller. "Population Characteristics of the Whooping Crane, *Grus americana*." *Canadian Journal of Zoology* 61 (1983): 2768–76.

Blankenship, D. R. "Aerial Search for Whooping Cranes along the Northeastern Mexican Coast." *American Birds* 27(1) (1973): 16.

Boyce, M. S., and R. S. Miller. "Ten-year Periodicity in Whooping Crane Census." *The Auk* 102 (1985): 659–60.

Butler, M. J., G. Harris, and B. N. Strobel. "Influence of Whooping Crane Population Dynamics on Its Recovery and Management." *Biological Conservation* 162 (2013): 89–99. http://dx.doi.org./10.1016/j.biocon.2013.04.003. Accessed May 14, 2015.

Butler, M. J., K. L. Metzger, and G. Harris. "Whooping Crane Demographic Responses to Winter Drought Focus Conservation Strategies." *Biological Conservation* 179 (2014): 72–85. http://dx.doi.org/10.1016/ j.biocon.2014.08.021. Accessed May 14, 2015.

Canadian Wildlife Service and U.S. Fish and Wildlife Service. *International Recovery Plan for the Whooping Crane*. Ottawa: Recovery of Nationally Endangered Wildlife (RENEW) and Albuquerque, NM: U.S. Fish and Wildlife Service, 2005.

Cannon, J. R., B. W. Johns, and T. V. Stehn. "Egg Collection and Recruitment of Young of the Year in the Aransas/Wood Buffalo Population of Whooping Cranes." *Proceedings North American Crane Workshop* 8 (2001): 11–16.

Chavez-Ramirez, F. "Food Availability, Foraging Ecology, and Energetics of Whooping Cranes Wintering in Texas." PhD diss., Texas A&M University, 1996.

———. "Report of Plaintiff's Responsive Expert Felipe Chavez-Ramirez, September 2, 2011." U.S. District Court for the Southern

District of Texas, Corpus Christi Division. Civil action No. 2:10-cv-00075, 2011.

Cornell Lab of Ornithology. "Whooping Crane Population Hits Historic High in 2018." https://www.allaboutbirds.org/whooping-crane-population-hits-historic-high-in-2018/, 2019. Accessed August 8, 2019.

COSEWIC. COSEWIC *Assessment and Status Report on the Whooping Crane* Grus americana *in Canada*. Committee on the Status of Endangered Wildlife in Canada, Ottawa, 2010: vii, 36.

Cracraft, J. "The Whooping Crane from the Lower Pleistocene of Arizona." *The Wilson Bulletin* 80(4) (1968): 490.

Craven, E. "The Status of the Whooping Crane on the Aransas Refuge, Texas." *The Condor* 48(1) (1946): 37–39.

Dambach, C. A. "Status of the Whooping Crane." *The Wilson Bulletin* 56(3) (1944): 180.

Doughty, R. W. *Return of the Whooping Crane*. Austin: University of Texas Press, 1989: x, 182.

Drewien, R. C., J. Tautin, M. L. Courville, and G. M. Gomez. "Whooping Cranes Breeding at White Lake, Louisiana, 1939: Observations by John J. Lynch, U.S. Bureau of Biological Survey." *Proceedings of the North American Crane Workshop* 8 (2001): 24–30.

Evans, D. E. and M. R. Waring. "Historical Loss of Whooping Crane Habitat on the Aransas National Wildlife Refuge." *Proceedings of the Annual Conference, Southeast Association of Fish and Wildlife Agencies* 47 (1993): 370–77.

Executive Order 7784. *Establishing the Aransas Migratory Waterfowl Refuge, Texas*, December 31, 1937. *Federal Register*, January 5, 1938, p. 347. Washington, DC: U.S. Government Printing Office.

Faanes, C. A., D. H. Johnson, and G. R. Lingle. "Characteristics of Whooping Crane Roost Sites in the Platte River." *Proceedings of the North American Crane Workshop* 6 (1992): 90–94.

Forbes, S., and D. W. Mock. "A Tale of Two Strategies: Life-History Aspects of Family Strife." *The Condor* 102 (2000): 23–34.

Gil-Weir, K. C. "Whooping Crane (*Grus americana*) Demography and Environmental Factors in a Population Growth Simulation Model." PhD diss., Texas A&M University, 2006: xi, 159.

Gil-Weir, K. C., F. Chavez-Ramirez, B. Johns, T. Stehn, and R. Silva.

"An Individual Whooping Crane's Family History." *Proceedings of the North American Crane Workshop* 11 (2010): 201.

Gil-Weir, K.C., W. E. Grant, R. D. Slack, H. H. Wang, and M. Fujiwara. "Demography and Population Trends of Whooping Cranes." *J. Field Ornithology* 83(1) (2012): 1–10.

Gill, S. "Sea-level Measurement, Determination, and Application in the Gulf of Mexico." Sea Level Rise 2010 Conference. Corpus Christi: Texas A&M University–Corpus Christi, Harte Research Institute, 2010: 42.

Gomez, G. M. "Whooping Cranes in Southwest Louisiana: History and Human Attitudes." *Proceedings of the North American Crane Workshop* 6 (1992): 19–23.

Gomez, G. M., R. C. Drewien, and M. L. Courville. "Historical Notes on Whooping Cranes at White Lake, Louisiana: The John J. Lynch interviews, 1947–1948." *Proceedings of the North American Crane Workshop* 9 (2005): 111–16.

Greer, D. M. "Blue Crab Population Ecology and Use by Foraging Whooping Cranes on the Texas Gulf Coast." PhD diss., Texas A&M University, 2010: xxxi, 294.

Guillen, G., and J. Oakley. *Bycatch Mortality and Critical Life History Parameters of the Texas Diamondback Terrapin.* Environmental Institute of Houston, University of Houston–Clear Lake, and Texas Parks and Wildlife Department, Dickinson, TX: Dickinson Marine Laboratory, 2013: 151.

Howe, M. A. *Migration of Radio-marked Whooping Cranes from the Aransas–Wood Buffalo Population: Patterns of Habitat Use, Behavior and Survival.* U.S. Dept. of the Interior, Fish and Wildlife Service, Technical Report 21 (1981): 33.

Hunt, H. E. "The Effects of Burning and Grazing on Habitat Use by Whooping Cranes and Sandhill Cranes on the Aransas National Wildlife Refuge, Texas." PhD diss., Texas A&M University, 1987: xvi, 173.

International Crane Foundation. "Whooping Crane Eastern Population Update — April 2019." https://www.savingcranes.org/whooping-crane-eastern-population-update-april-2019/, 2019. Accessed May 22, 2019.

Jenniges, J. J., and M. M. Peyton. "Management of Lands along the

Platte River from Elm Creek to Lexington, Nebraska as Crane Habitat." *Proceedings of the North American Crane Workshop* 10 (2008): 76–85.

Johns, B. W. "Preliminary Identification of Whooping Crane Staging Areas in Prairie Canada." *North American Crane Workshop Proceedings*, Paper 291. http://digitalcommons.unl.edu/nacwgproc/291, 1992.

Johns, B. W., J. Goossen, E. Kuyt, and L. Craig-Moore. "Philopatry and Dispersal in Whooping Cranes." *Proceedings of the North American Crane Workshop* 9 (2005): 117–25.

Johns, B. W., E. J. Woodsworth and E. A. Driver. "Habitat Use by Whooping Cranes in Saskatchewan." *Proceedings of the North American Crane Workshop* 7 (1997): 123–31.

Johnsgard, P. A. *Cranes of the World.* Chapter 2: Individualistic and Social Behavior. Papers in the Biological Sciences. Lincoln: University of Nebraska Press, 1983: 11–24.

Kuyt, E. *Aerial Radio-tracking of Whooping Cranes Migrating between Wood Buffalo National Park and Aransas National Wildlife Refuge, 1981–84.* Canadian Wildlife Service, Occasional Paper 74 (1992): 53.

———. "Whooping Crane." In *Proceedings of the Workshop on Endangered Species in the Prairie Provinces.* Edited by G. L. Holroyd, W. B. McGillivray, P. H. R. Stepney, D. M. Ealey, G. C. Trottier, and K. E. Eberhart. Alberta Culture, Historical Resources Division. Occasional Paper No. 9 (1987): 229–32.

Lewis, T. E., and R. D. Black. "Whooping Cranes and Human Disturbance: An Historical Perspective and Literature Review." *Proceedings of the North American Crane Workshop* 10 (2008): 3–6.

Lockwood, M. W., and B. Freeman. *The TOS Handbook of Texas Birds.* College Station: Texas A&M University Press, 2004: xxvii, 261.

Louisiana Department of Wildlife and Fisheries. "Whooping Cranes." http://www.wlf.louisiana.gov/wildlife/whooping-cranes, 2019. Accessed May 22, 2019.

Lyons, J. R. "FDR and Environmental Leadership." In *FDR and the Environment.* Edited by D. B. Woolner and H. L. Henderson. London: Palgrave Macmillan, 2009: ix, 265.

Matthiessen, P. *The Birds of Heaven.* New York: North Point Press,

2001: xv, 347.

McAlister, W. H., and M. K. McAlister. *Aransas: A Naturalist's Guide.* Austin: University of Texas Press, 1995: viii, 392.

McIlhenny, E. A. "Major Changes in the Bird Life of Southern Louisiana during Sixty Years." *The Auk* 60 (1943): 541–49.

——. "Whooping Crane in Louisiana." *The Auk* 55(4) (1938): 670.

McNulty, F. *The Whooping Crane.* London: Longmans, Green and Co., 1967: 190

Mershon, W. B. "Whooping Crane in Saskatchewan." *The Auk* 45 (1928): 202–3.

Miller, R. S. "The Brood Size of Cranes." *The Wilson Bulletin* 85(4) (1973): 436–41.

Mills, H. B., and F. C. Bellrose. "Whooping Crane in the Mid-West." *Auk* 76 (1959): 234–35.

National Wildlife Refuge System. http://fws.gov/refuges/news/RefugesCelebrate75Years_12132010.html. Accessed March 15, 2015.

Nelson, E. W. "The Whooping Crane Continues to Visit Louisiana." *Condor* 31(4) (1929): 146–47.

Nelson, J. T., R. D. Slack, and G. F. Gee. "Nutritional Value of Winter Foods for Whooping Cranes." *The Wilson Bulletin* 108(4) (1996): 728–39.

Nesbitt, S. A. "Do We Need Such Rare Birds?" *Proceedings of the North American Crane Workshop* 10 (2006): 1–2.

Nesbitt, S. A., and M. J. Folk. "Reintroduction of the Whooping Crane in Florida." *North American Birds* 54 (2000): 248.

Nigge, K. *Whooping Crane: Images from the Wild.* College Station: Texas A&M University Press, 2010: vii, 217.

Novakowski, N. S. *Whooping Crane Population Dynamics on the Nesting Grounds, Wood Buffalo National Park, Northwest Territories, Canada.* Ottawa: Canadian Wildlife Service Report Series Number 1 (1966): 21.

Oberholser, H. C. *The Bird Life of Texas*, vol. 1. Edited by E. B. Kincaid Jr. Austin: University of Texas Press, 1974: xxviii, 530.

Olson, S. L. "A Whooping Crane from the Pleistocene of North Florida." *The Condor* 74(3) (1972): 341.

Pearse, A., D. Baasch, M. Bidwell, M. Harner, and B. Strobel. "Remote Tracking of Aransas–Wood Buffalo Whooping Cranes 2012 Breeding Season and Fall Migration Update." U.S. Dept. of the Interior, U.S. Geological Survey, 2013: 6.

Pearse, A., D. A. Brandt, W. C. Harrell, K. L. Metzger, D. M. Baasch, and T. J. Hefley. "Whooping Crane Stopover Site Use Intensity within the Great Plains." U.S. Geological Survey Open-File Report 2015 — 1166. http://dx.doi.org/10.3133/ofr20151166, 2015. Accessed April 23, 2014.

Pearson, T. G. "Whooping Cranes in Texas." *The Auk* 34 (1922): 412–13.

Slack, R. D, W. E. Grant, S. E. Davis III, T. M. Swannack, J. Wozniak, D. M. Greer, and A. G. Snelgrove. *Linking Freshwater Inflows and Marsh Community Dynamics in San Antonio Bay to Whooping Cranes.* Final Report: San Antonio Guadalupe Estuarine System. College Station: Texas A&M AgriLIFE, 2009: xiv, 173.

Sotiropoulos, M. A. "Analysis of Food Pond Webs in the Whooping Crane Nesting Area, Wood Buffalo National Park." Master's thesis, University of Alberta, 2002: 100.

Sprunt IV, A. "In Memoriam: Robert Porter Allen." *The Auk* 86 (1969): 26–34.

Stahlecker, D. W. "Availability of Stopover Habitat for Migrant Whooping Cranes in Nebraska." *Proceedings of the North American Crane Workshop* 7 (1997): 132–40.

Stehn, T. "Behavior of Whooping Cranes during Initiation of Migration." *Proceedings of the North American Crane Workshop* 6 (1992): 102–5.

——. "Comments on SAGES [San Antonio Guadalupe Estuarine System] Final Report." Submitted to U.S. District Court for the Southern District of Texas, Corpus Christi Division. Civil action No. 2:10-cv-00075, 2009: 72.

——. "Migration of Radio-monitored Cranes from Aransas to Wood Buffalo: Family No. 19/83." Unpublished files. U.S. Fish and Wildlife Service, Aransas National Wildlife Service, 1984: 30.

——. "Pair Formation by Color-marked Whooping Cranes on the Wintering Grounds." *Proceedings of the North American Crane Workshop* 7 (1997): 24–28.

——. "Radio Tracking Whoopers across a Continent." *Fish and Wildlife News.* June–July 1984: 6–8.

——. "Tailing the Whoopers." *Texas Parks and Wildlife Magazine.*

March 1985: 18–21.

———. "Unusual Movements and Behaviors of Color-banded Whooping Cranes during Winter." *Proceedings of the North American Crane Workshop* 6 (1992): 95–101.

Stehn, T., & F. Prieto. "Changes in Winter Whooping Crane Territories and Range 1950–2006." *Proceedings of the North American Crane Workshop* 11 (2010): 40–56.

Stehn, T., & T. Wassenich. "Whooping Crane Collisions with Power Lines: An Issue Paper." *Proceedings of the North American Crane Workshop* 10 (2008): 25–36.

Stevenson, J. O. "Whooping Cranes in Texas in Summer." *The Condor* 44 (1942): 40–41.

Stevenson, J. O., & R. E. Griffith. "Winter Life of the Whooping Crane." *The Condor* 48 (1946): 160–78.

Taylor, L. N., L. P. Ketzler, D. Rousseau, B. N. Strobel, K. L. Metzger, and M. J. Butler. "Observations of Whooping Cranes during Winter Aerial Surveys: 1950–2011." Aransas National Wildlife Refuge, U.S. Fish and Wildlife Service, 2015. http://dx.doi.org/10.7944/W3RP4B. Accessed July 7, 2015.

Texas Department of Transportation. *The Gulf Intracoastal Waterway.* Legislative Report — 83rd Legislature. Austin: Texas Department of Transportation. n.d.: 16.

Texas History. "Whooping Cranes May Be Increasing." *Palacios Beacon.* December 3, 1953. http://texashistory.unt.edu/search/?q=whooping+crane&start=100&t=fulltext. Accessed August 20, 2015.

Timoney, K. "The Habitat of Nesting Whooping Cranes." *Biological Conservation* 89 (1999): 189–97.

Tufty, B. "28 Whooping Cranes left." *The Science New-Letter* 83(16) (1963): 245.

U.S. Department of Agriculture. "Texas Ranch Now a Waterfowl Refuge." Press release, April 25, 1939. Washington, DC: 2.

U.S. Fish and Wildlife Service. *White-tailed Deer, Feral Hog and Waterfowl Hunt Plan.* Austwell, TX: Aransas National Wildlife Refuge, November 2012.

U.S. Fish and Wildlife Service. *Whooping Crane* (Grus americana) *5-Year Review: Summary and Evaluation.* Austwell, TX: Aransas National Wildlife Refuge and Corpus Christi Ecological Field

Services Office, 2011: 44.

U.S. Fish and Wildlife Service. "Whooping Crane Update, Aransas National Wildlife Refuge." Press release, April 12, 2012: 1.

U.S. Geological Survey. "Remote Tracking of Aransas–Wood Buffalo Whooping Cranes 2014–2015." Press release, June 2015: 9.

U.S. Geological Survey. "Remote Tracking of the Aransas–Wood Buffalo Whooping Cranes, 2012 Breeding Season and Fall Migration Updates." Press release, February 2013: 6.

Walkinshaw, L. H. "Attentiveness of Cranes at their Nest." *The Auk* 82 (1965): 465–76.

Ward, G. H. "The Blue Crab: A Survey with Application to San Antonio Bay." *Biological Study of San Antonio Bay*. Austin: Texas Water Development Board-University of Texas, 2012: xiv, 210.

Westwood, C. M., and F. Chavez-Ramirez. "Patterns of Food Use of Wintering Whooping Cranes on the Texas Coast." *Proceedings of the North American Crane Workshop* 9 (2005): 133–40.

White, J. L. *Status of the Whooping Crane (Grus americana) in Alberta*. Alberta Wildlife Status Report No. 34. Fisheries and Wildlife Management Division. Edmonton: Alberta Environment and Alberta Conservation Association, 2001: 22.

Whooping Crane Conservation Association. "Whooping Cranes from Despair to Hope to Progress." http://whoopingcrane.com/whooping-cranes-from-despair-to-hope-to-progress, December 23, 2012; for article from September 20, 1954 *Sports Illustrated*. Accessed April 26, 2015.

Wilson, L. "Analysis of the Science: The Whooping Crane Decision, the Aransas Project vs. Shaw." Texas Public Policy Foundation, 2013: 24.

Wright, G. D., M. J. Harner, and J. D. Chambers. "Unusual Wintering Distribution and Migratory Behavior of the Whooping Crane (*Grus americana*) in 2011–2012." *Wilson Journal of Ornithology* 126(1) (2014): 115–20.

CHAPTER 4

Adair, S. E., J. L. Moore, and W. H. Kiel Jr. "Wintering Diving Duck Use of Coastal Ponds: An Analysis of Alternative Hypotheses." *The Journal of Wildlife Management* 60(1) (1996): 83–93.

Alperin, L. M. *History of the Gulf Intracoastal Waterway.* Navigation History NWS-83–89. National Waterways Study, U.S. Army Engineer Water Resources Support Center, Institute for Water Resources, 1983.

Andersen, H. C. *Hans Andersen's Fairy Tales.* Translation by Naomi Lewis. London: Penguin Books, 2010: xxii, 182.

Austin, J., T. Buhl, G. R. Guntenspergen, W. Norling, and H. T. Sklebar. "Duck Populations as Indicators of Landscape Condition in the Prairie Pothole Region." USGS *Northern Prairie Wildlife Research Center.* Paper 8, 2001. http://digitalcommons.unl.edu/usgsnpwrc/8.

Bailey, R. O., and R. D. Titman. "Habitat Use and Feeding Ecology of Post-breeding Redheads." *The Journal of Wildlife Management* 48(4) (1984): 1144–55.

Baldassarre, G. A. "Field-feeding Ecology of Waterfowl Wintering on the Southern High Plains of Texas." PhD diss., Texas Tech University, 1982: ix, 52.

Ballard, B. M., J. E. Thompson, M. J. Petrie, M. Chekett, and D. J. Hewitt. "Diet and Nutrition of Northern Pintails Wintering along the Southern Coast of Texas." *The Journal of Wildlife Management* 68(2) (2004): 371–82.

Beedy, E. C., and B. E. Deuel. "Species Account: Redhead." In "California Bird Species of Special Concern: A Ranked Assessment of Species, Subspecies, and Distinct Populations of Birds of Immediate Conservation Concern in California." *Studies of Western Birds,* 85–90. Edited by W. D. Shuford, and T. Garaldi. Camarillo, CA: Western Field Ornithologists and Sacramento: California Department of Fish and Game, 2008.

Bellrose, F. C. *Ducks, Geese & Swans of North America.* Harrisburg PA: Stackpole Books, 1976: 543.

Bolen, E. G. "Waterfowl Management: Yesterday and Tomorrow." *The Journal of Wildlife Management* 64(2) (2000): 323–35.

Brewster, W. G., J. M. Gates, and L. D. Flake. "Breeding Waterfowl Populations and Their Distribution in South Dakota." *The Journal of Wildlife Management* 40(1) (1976): 50–56.

Buskey, E. J., H. Liu, C. Collumb, and J. G. F. Bersano. "The Decline and Recovery of a Persistent Texas Brown Tide Algal Bloom in the Laguna Madre (Texas, USA)." *Estuaries* 24(3) (2001): 337–46.

Buskey, E. J., S. Stewart, J. Peterson, and C. Collumb. *Current Status and Historical Trends of Brown Tide and Red Tide Phytoplankton Blooms in the Corpus Christi Bay National Estuary Program Study Area.* Port Aransas, TX: Marine Science Institute, University of Texas at Austin. Corpus Christi Bay National Estuary Program Report No. CCBNEP-07, 1996: xiv, 65.

Cornelius, S. E. "Food and Resource Utilization by Wintering Redheads on Lower Laguna Madre." *The Journal of Wildlife Management* 41(3) (1977): 374–85.

———. "Wetland Salinity and Salt Gland Size in the Redhead *Aythya americana.*" *The Auk* 99(4) (1982): 774–78.

Custer, C. M. "Life History Traits and Habitat Needs of the Redhead." Fish and Wildlife Leaflet 13.1.11. In *Waterfowl Management Handbook.* U.S. Fish and Wildlife Service, 1993: 1–7. http://www.nwrc. usgs.gov/wdb/pub/wmh/13_1_11.pdf. Accessed March 11, 2015.

Custer, C. M., T. W. Custer, and P. J. Zwank. "Migration Chronology and Distribution of Redheads on the Lower Laguna Madre, Texas." *The Southwestern Naturalist* 42(1) (1997): 40–51.

Dahl, T. E. *Status and Trends of Prairie Wetlands in the United States 1997 to 2009.* Washington, DC: U.S. Department of the Interior; Fish and Wildlife Service, Ecological Services, 2014.

Dahl, T. E., and S. M. Stedman. *Status and Trends of Wetlands in the Coastal Watersheds of the Coterminous United States 2004 to 2009.* U.S. Department of the Interior, Fish and Wildlife Service and National Oceanic and Atmospheric Administration, National Marine Fisheries Service, 2013: 46.

Doherty, K. E., A. J. Ryba, C. L. Stemler, N. D. Niemuth, and W. A. Meeks. "Conservation Planning in an Era of Change: State of the U.S. Prairie Pothole Region." *Wildlife Society Bulletin* 37(3) (2013): 546–63. Published online doi: 10.1002/wsb.284.

Dunton, K., W. Pulich, and T. Mutchler. *A Seagrass Monitoring Program for Texas Coastal Waters: Multiscale Integration of Landscape Features with Plant and Water Quality Indicators.* Contract No. 0627. Corpus Christi: Coastal Bend Bays and Estuaries Program, 2010: 39.

Dunton, K,. and S. V. Schonberg. "Assessment of Propeller Scarring in Seagrass Beds of the South Texas Coast." *Journal of Coastal Research* 37 (2002): 100–110.

Erftemeijer, P. L. A., and R. R. Robin Lewis III. "Environmental Impacts of Dredging on Seagrasses: A Review." *Marine Pollution Bulletin* 52 (2006): 1553–72.

Esslinger, C. G., and B. C. Wilson. *North American Waterfowl Management Plan, Gulf Coast Joint Venture: Chenier Plain Initiative.* Albuquerque, NM: North American Waterfowl Management Plan, 2001: 28, appendix. (Revised 2003.)

Fulbright, T. E., and Fred C. Bryant. "The Wild Horse Desert: Climate and Ecology." In *Ranch Management: Integrating Cattle, Wildlife, and Range,* 35–58. Edited by C. A. Forgason, F. C. Bryant, and P. C. Genho. King Ranch Institute, 2003.

Fulbright, T. E., D. D. Diamond, J. Rappole, and J. Norwine. "Coastal Sand Plain of Southern Texas." *Rangelands* 12(6) (1990): 337–40.

Garcia, C. P. *Captain Alonso Pineda and the Exploration of the Texas Coast and the Gulf of Mexico.* Austin: The San Felipe Press, 1982: 62.

Greenwood, R. J, A. B. Sargeant, D. H. Johnson, L. M. Cowardin, and T. L. Shaffer. "Factors Associated with Duck Nest Success in the Prairie Pothole Region of Canada." USGS Northern Prairie Wildlife Research Center. Paper 243. *Wildlife Monographs* 128 (1995): 1–57. Accessed at http://digitalcommons.unl.edu/usgsnpwrc/243.

Gutierrez, M. A., A. A. Cardona, and D. L. Smee. "Growth Patterns of Shoal Grass *Halodule wrightii* and Manatee Grass *Syringodium filiforme* in the Western Gulf of Mexico." *Gulf and Caribbean Research* 22 (2010): 71–75.

Handley, L., D. Altsman, and R. DeMay, eds. *Seagrass Status and Trends in the Northern Gulf of Mexico: 1940–2002.* U.S. Geological Survey Scientific Investigations Report 2006-5287 and U.S. Environmental Protection Agency 855-R-04-003, 2007: 267.

Haramis, G. M. "Redhead *Aythya americana.*" In *Habitat Requirements for Chesapeake Bay Living Resources.* Edited by S. L Funderburk, S. J. Jordan, J. A. Mihursky, and D. Riley. Annapolis, MD: Patuxent Wildlife Research Center, 1991: 18-1 to 18-10. http://www.dnr.state.md.us/irc/docs/00000260_18.pdf. Accessed May 9, 2015.

Haywood, K. McD. "The Laguna Madre of Texas: A History and Analysis of the Spatial Understanding and Cultural Constructions of its Fisheries." PhD diss., University of Texas, 2003: xiii, 255.

Hedgpeth, J. W. "The Laguna Madre of Texas." *North American Wildlife*

Conference 12 (1947): 364–80.

Hochbaum, H. A. *Travels and Traditions of Waterfowl.* Minneapolis: University of Minnesota Press, 1955: x, 300.

Iverson, R. L., and H. F. Bittaker. "Seagrass Distribution and Abundance in Eastern Gulf of Mexico Coastal Waters." *Estuarine, Coastal and Shelf Science* 22 (1986): 577–602.

James, J. D. "Utilization of Shoalgrass and Nutritional Ecology of Wintering Redheads in the Laguna Madre of Texas." PhD diss., Texas A&M University, 2006: xvii, 117.

Johnsgard, Paul A. "Waterfowl of North America: Waterfowl Distributions and Migrations in North America." In *Waterfowl of North America*, revised ed. Paper 5, 2010: 10. http://digitalcommons.unl. edu/biosciwaterfowlna/5. Accessed June 3, 2015.

Joyner, D. E. "Parasitic Egg Laying in Redheads and Ruddy Ducks in Utah: Incidence and Success." *The Auk* 100 (1983): 717–25.

Kahara, S. N., R. M. Mockler, K. F. Higgins, S. R. Chipps, and R. R. Johnson. "Spatiotemporal Patterns of Wetland Occurrence in the Prairie Pothole Region of Eastern South Dakota." *Wetlands* 29(2) (2009): 678–89.

Kaldy, J. E., K. H. Dunton, J. L. Kowalski, and K. Lee. "Factors Controlling Seagrass Revegetation onto Dredged Material Deposits: A Case Study in Lower Laguna Madre, Texas." *Journal of Coastal Research* 20(1) (2004): 292–300.

Kaldy, J. E., C. P. Onuf, P. M. Eldridge, and L. A. Cifuentes. "Carbon Budget for a Subtropical Seagrass Dominated Coastal Lagoon: How Important Are Seagrasses to Total Ecosystem Net Primary Production?" *Estuaries* 25(4) (2002) Part A: 528–39.

Kantrud, H. A. and R. E. Stewart. "Use of Natural Basin Wetlands by Breeding Waterfowl in North Dakota." *The Journal of Wildlife Management* 41(2) (1977): 243–53.

Kenow, K. P., and D. H. Rusch. "Food Habits of Redheads at the Horicon Marsh, Wisconsin." *Journal of Field Ornithology* 67(4) (1996): 649–59.

Klett, A. T., T. L. Shaffer, and D. H. Johnson. "Duck Nest Success in the Prairie Pothole Region." *The Journal of Wildlife Management* 52(3) (1988): 431–40.

Klimstra, J. D., and P. I. Padding. *Atlantic Flyway Harvest and Popula-*

tion Survey Data Book. Laurel, MD: U.S. Fish and Wildlife Service, 2013: 94.

Klopfleisch, R., C. Müller, U. Polster, J-P. Hildebrandt, and J. Teifke. "Granulomatous Inflammation of Salt Glands in Duckling (*Anas platyrhynchos*) Associated with Intralesional Gram-negative Bacteria." *Avian Pathology* 34(3) (2005): 233–37.

Kowalski, J. L. "Production of the Subtropical Seagrass *Halodule wrightii* Aschers, in Lower Laguna Madre." Master's thesis, University of Texas–Pan American, 1999: xiii, 105.

Kuvlesky, W. P., Jr., L. A. Brennan, M. L. Morrison, K. K. Boydston, B. M. Ballard, and F. C. Bryant. "Wind Energy Development and Wildlife Conservation: Challenges and Opportunities." *The Journal of Wildlife Management* 71(8) (2007): 2487–98.

Lawson, R. W. *Frontier Naturalist: Jean Louis Berlandier and the Exploration of Northern Mexico and Texas.* Albuquerque: University of New Mexico Press, 2012: xxi, 262.

Lockwood, M. W. and B. Freeman. *The TOS Handbook of Texas Birds.* College Station: Texas A&M University Press, 2004: xxvii, 261.

Lokemoen, J. T. "Breeding Ecology of the Redhead Duck in Western Montana." *The Journal of Wildlife Management* 30(4) (1966): 668–81.

Loss, S. R., T. Will, and P. Mara. "Estimates of Bird Collision Mortality at Wind Facilities in the Contiguous United States." *Biological Conservation* 168 (2013): 201–9.

Maryland Department of Natural Resources. 2013 "Mid-winter Waterfowl Survey Results." htt://news.maryland.gov/dnr/2013/02/0 5/2013-midwinter-waterfow-survey-results-are-in/, 2014. Accessed June 19, 2014.

McKee, D. A. *Fishes of the Texas Laguna Madre.* College Station: Texas A&M University Press, 2008: xvii, 203.

McMahan, C. A. "Biomass and Salinity Tolerance of Shoalgrass and Manateegrass in Lower Laguna Madre, Texas." *The Journal of Wildlife Management* 32(3) (1968): 501–6.

——. "Food Habits of Ducks Wintering on Laguna Madre, Texas." *The Journal of Wildlife Management* 34(4) (1970): 946–49.

McMahan, C. A., and R. L. Fritz. "Mortality to Ducks from Trotlines in Lower Laguna Madre, Texas." *The Journal of Wildlife Management* 31(4) (1967): 783–87.

Michot, T. C. "Carrying Capacity of Seagrass Beds Predicted For Redheads Wintering in Chandeleur Sound, Louisiana, USA." In *Effect of Habitat Loss and Change on Waterbirds*. Edited by J. D. Goss-Custard, R. Rufino, and A. Luis. Institute of Terrestrial Ecology Symposium No. 30, Wetlands International Publication No. 42. Proceedings of the 10th International Waterfowl Ecology Symposium, University of Aveiro, Portugal, September 18–21, 1995: 93–102.

———. "Comparison of Wintering Redhead Populations in Four Gulf of Mexico Seagrass Beds." In *Limnology and Aquatic Birds: Monitoring, Modelling and Management*. Edited by F. A. Comin, J. A. Herrera-Silveira, and J. Ramirez-Ramirez, 243–60. Second International Symposium on Limnology and Aquatic Birds. Merida: Universidad Autonoma de Yucatan, 2000.

Michot, T. C., M. C. Garvin, and E. H. Weidner. "Survey for Blood Parasites in Redheads (*Aythya americana*) Wintering at the Chandeleur Islands, Louisiana." *Journal of Wildlife Diseases* 31(1) (1995): 90–92.

Michot, T. C., and A. J. Nault. "Diet Differences in Redheads from Nearshore and Offshore Zones in Louisiana." *The Journal of Wildlife Management* 57(2) (1993): 238–44.

Michot, T. C., M. C. Woodin, and A. J. Nault. "Food Habits of Redheads (*Aythya americana*) Wintering in Seagrass Beds of Coastal Louisiana and Texas, USA." *Acta Zoologica Academiae Scientarum Hungaricae* 54 (Suppl. 1) (2008): 239–50.

Mitchell, C. A. "Water Depth Predicts Redhead Distribution in the lower Laguna Madre, Texas." *Wildlife Society Bulletin* 20(4) (1992): 420–24.

Mitchell, C. A., T. W. Custer, and P. J. Zwank. "Herbivory on Shoalgrass by Wintering Redheads in Texas." *The Journal of Wildlife Management* 58(1) (1994): 131–41.

Mitchell, C. A., T. W. Custer, and P. J. Zwank. "Redhead Duck Behavior on Lower Laguna Madre and Adjacent Ponds of Southern Texas." *The Southwestern Naturalist* 37(1) (1992): 65–72.

Montagna, P. A. and R. D. Kalke. "Ecology of Infaunal Mollusca in South Texas Estuaries." *American Malacological Bulletin* 11(2) (1995): 163–75.

Muehl, G. T. "Distribution and Abundance of Water Birds and Wetlands in Coastal Texas." Master's thesis, Texas A&M University–Kingsville, 1994: xii, 130.

National Oceanic and Atmospheric Administration. "'The Valleys' Ceaseless Wind?" Brownsville, TX: National Weather Service Forecast Office, 2008. http://www.srh.noaa.gov/bro?n=2008event_ceaselesswind. Accessed July 11, 2015.

Oberholser, H. C., and E. B. Kinkaid, Jr., eds. *The Bird Life of Texas*, vol. 1. Austin: University of Texas Press, 1974: xxvii, 530.

Onuf, C. P. "Laguna Madre." In *Seagrass Status and Trends in the Northern Gulf of Mexico: 1940–2002*. Edited by L. Handley, D. Altsman, and R. DeMay, 29–40. U.S. Geological Survey Scientific Investigations Report 2006-5287 and U.S. Environmental Protection Agency 855-R-04-003, 2007: 267.

———. "Seagrasses, Dredging, and Light in Laguna Madre, Texas, USA." *Estuarine Coastal and Shelf Science* 39(1) (1994): 75–91.

Orth, R. J. "Distribution and Abundance of Submerged Aquatic Vegetation in Chesapeake Bay: An Historical Perspective." *Estuaries* 7(48) (1984): 531–40.

Orth, R. J., T. B. J. Carruthers, W. C. Dennison, C. M. Duarte, J. W. Fourqurean, K. L. Heck Jr., A. R. Hughes, G. A. Kendrick, W. J. Kenworthy, S. Olyarnik, F. T. Short, M. Waycott, and S. L. Williams. "A Global Crisis for Seagrass Ecosystems. *BioScience* 56(12) (2006): 987–96.

Orth, R. J., and K. A. Moore. "Chesapeake Bay: An Unprecedented Decline in Submerged Aquatic Vegetation." *Science, New Series* 222(4619) (1983): 51–53.

Phillips, J. C. "Red-Head." In *A Natural History of the Ducks*, vol. III, 163–81, vol. IV. New York: Dover Publications Inc., 1986: ix, 383 (vol. III), ix, 489 (vol. IV).

Pilkey, O. H., J. A. G. Cooper, and D. A. Lewis. "Global Distribution and Geomorphology of Fetch-limited Barrier Islands." *Journal of Coastal Research* 25(4) (2009): 819–37, 925–29.

Pulich, W. M. "Texas Coastal Bend" In *Seagrass Status and Trends in the Northern Gulf of Mexico: 1940–2002*. Edited by L. Handley, D. Altsman, and R. DeMay, 41–59. U.S. Geological Survey Scientific Investigations Report 2006-5287 and U.S. Environmental Protection Agency 855-R-04-003, 2007: 267.

Pulich, W. M., B. Hardegree, A. Kopecky, S. Schwelling, C. Onuf, and K. Dunton. *Texas Seagrass Monitoring Program Strategic Plan.* Austin: Texas Parks and Wildlife Department, Texas Commission on Environmental Quality, and Texas General Land Office, 2003.

Pulich, W. M., and C. Onuf. "Statewide Summary for Texas." In *Seagrass Status and Trends in the Northern Gulf of Mexico: 1940–2002.* Edited by L. Handley, D. Altsman, and R. DeMay, 7–16. U.S. Geological Survey Scientific Investigations Report 2006–5287 and U.S. Environmental Protection Agency 855-R-04-003, 2007: 267.

Pulich, W. M., and W. A. White. "Decline of Submerged Vegetation in the Galveston Bay System: Chronology and Relationships to Physical Processes." *Journal of Coastal Research* 7(4) (1991): 1125–38.

Quammen, M. L., and C. P. Onuf. "Laguna Madre: Seagrass Changes Continue Decades after Salinity Reduction." *Estuaries* 16(2) (1993): 302–10.

Ralph, P. J., M. J. Durako, S. Enríquez, C. J. Collier, and M. A. Doblin. "Impact of Light Limitation on Seagrasses." *Journal of Experimental Marine Biology and Ecology* 350 (2007): 176–93.

Ray, J. D., B. D. Sullivan, and H. W. Miller. "Breeding Ducks and Their Habitats in the High Plains of Texas." *The Southwestern Naturalist* 48(2) (2003): 241–48.

Reynolds, R. E., C. R. Loesch, B. Wangler, and T. L. Shaffer. *Waterfowl Response to the Conservation Reserve Program and Swampbuster Provision in the Prairie Pothole Region, 1992–2004.* Bismarck, ND: U.S. Fish and Wildlife Service; Jamestown, ND: U.S. Geologic Survey, 2007: 90. RFA 05-IA-04000000-N34.

Rhodes, M. J. "Redheads Breeding in the Texas Panhandle." *The Southwestern Naturalist* 24(4) (1979): 691–92.

Rickner, J. A. "The Influence of Dredged Material Islands in Upper Laguna Madre, Texas on Selected Seagrasses and Macro-benthos." PhD diss., Texas A&M University, 1979: ix, 57.

Rio Grande, Rio Grande Estuary, and Lower Laguna Madre Basin and Bay Expert Science Team for the Lower Rio Grande Basin. *Environmental Flows Recommendations Report.* Final Submission to the Environmental Flows Advisory Group, Rio Grande, Rio Grande Estuary, and Lower Laguna Madre Basin and Bay Stakeholders

Committee, and Texas Commission on Environmental Quality, 2012: vii, 228.

Robertson, G. J., and F. Cooke. "Winter Philopatry in Migratory Waterfowl." *The Auk* 116(1) (1999): 20–34.

Sargeant, A. B., R. J. Greenwood, M. A. Sovada, and T. L. Shaffer. *Distribution and Abundance of Predators that Affect Duck Production — Prairie Pothole Region.* Washington, DC: U.S. Fish and Wildlife Service Resource Publication 194, 1993: iii, 98.

Sawyer, R. K. *Texas Market Hunting: Stories of Waterfowl, Game Laws and Outlaws.* College Station: Texas A&M University Press, 2013: xi, 184.

Sheridan, P. "Comparison of Restored and Natural Seagrass Beds near Corpus Christi, Texas." *Estuaries* 27(5) (2004): 781–92.

———. "Recovery of Floral and Faunal Communities after Placement of Dredged Material on Seagrasses in Laguna Madre, Texas." *Estuarine, Coastal and Shelf Science* 59 (2004): 441–58.

Short, F. T., and S. Wyllie-Echeverria. "Natural and Human-induced Disturbance of Seagrass." *Environmental Conservation* 23(1) (1996): 17–27.

Singleton, J. R. *Waterfowl Management in Texas.* Austin: Texas Parks and Wildlife Department, Bulletin No. 47 (1965): 65.

Smith, E. H. "Redheads and Other Wintering Waterfowl." In *The Laguna Madre of Texas and Tamaulipas.* Edited by J. W. Tunnell Jr. and F. W. Judd, 169–81. College Station: Texas A&M University Press, 2002: xxii, 346.

Spiller, K., and R. Blankinship. "Laguna Madre." Texas Parks and Wildlife. https://tpwd.texas.gov/fishboat/fish/didyouknow/lagunamadre.phtml, 2015. Accessed September 28, 2015.

Stevenson, J. C., and N. M. Confer. *Summary of Available Information on Chesapeake Bay Submerged Vegetation.* U.S. Fish and Wildlife Service, Office of Biological Services Report No. FWS/OBS-78-66, 1978: xxii, 335.

Stout, I. J., and G. W. Cornwell. "Non-hunting Mortality of Fledged North American Waterfowl." *The Journal of Wildlife Management* 40(4) (1976): 681–93.

Sugden, L. G. "Parasitism of Canvasback nests by Redheads." *Journal of Field Ornithology* 51(4) (1980): 361–64.

Texas Coast Ocean Observation Network (TCOON). http://www.tcoon.org/. Accessed July 14, 2015.

Texas Department of Transportation. *Gulf Intracoastal Waterway Legislative Report.* Austin: Texas Department of Transportation, 2016: 21. http://ftp.dot.state.tx.us/pub/txdot-info/tpp/giww/legislative-report-85.pdf. Accessed May 23, 2019.

Texas Department of Transportation. *Gulf Intracoastal Waterway.* Legislative Report — 83rd Legislature. Austin: Texas Department of Transportation. n.d.: 16.

Texas Department of Transportation. *Gulf Intracoastal Waterway 2005–2006 Legislative Report.* Austin: Texas Department of Transportation, 2006: 21.

Texas Department of Water Resources. *Laguna Madre Estuary: A Study of the Influence of Freshwater Inflows.* Austin: Texas Department of Water Resources Report No. LP-182, 1983: xxi, 264.

Texas Parks and Wildlife. *Seagrass Conservation Plan for Texas.* Austin: Texas Parks and Wildlife, 1999: 79.

Texas Parks and Wildlife. "Mid-winter Waterfowl Survey." https://tpwd.texas.gov/huntwild/wild/game_management/waterfowl/, 2018. Accessed February 20, 2018.

Tunnell, J. W., Jr. "The Environment." In *The Laguna Madre of Texas and Tamaulipas.* Edited by J. W. Tunnell Jr. and F. W. Judd, 73–84. College Station: Texas A&M University Press, 2002: xxii, 346.

———. "Geography, Climate, and Hydrography." In *The Laguna Madre of Texas and Tamaulipas.* Edited by J. W. Tunnell Jr. and F. W. Judd, 7–27. College Station: Texas A&M University Press, 2002: xxii, 346.

———. "Origin, Development, and Geology." In *The Laguna Madre of Texas and Tamaulipas.* Edited by J. W. Tunnell Jr. and F. W. Judd, 28–37. College Station: Texas A&M University Press, 2002: xxii, 346.

U.S. Department of the Interior, Environment Canada, and Ministry of Environment and Natural Resources, Mexico. *North American Waterfowl Management Plan — People Conserving Waterfowl and Wetlands,* 2012: xx, 48.

U.S. Fish and Wildlife Service. "Central Flyway Mid-winter Survey Results." Lakewood, CO: Division of Migratory Bird Management,

2013: 34.

U.S. Fish and Wildlife Service. "Mid-winter Waterfowl Survey." https://migbirdapps.fws.gov/mbdc/databases/mwi/statezone.asp, 2019. Accessed May 24, 2019.

U.S. Fish and Wildlife Service. *Waterfowl Population Status, 2018*. Washington, DC: U.S. Department of the Interior, 2018: 71.

U.S. Fish and Wildlife Service. *Waterfowl Population Status, 2013*. Washington, DC: U.S. Department of the Interior, 2013: 79.

U.S. Government Accountability Office. *Prairie Pothole Region: At the Current Pace of Acquisitions, the U.S. Fish and Wildlife Service Is Unlikely to Achieve Its Habitat Protection Goals for Migratory Birds.* Washington, DC. GAO-07-1093, 2007: 50.

Villareal, T.A., T. Chirichella, and E.J. Buskey. "Regional Distribution of the Texas Brown Tide (*Aureoumbra lagunensis*) in the Gulf of Mexico." In *Harmful Algae 2002.* Edited by K.A. Steidinger, J. H. Landsberg, C. R. Tomas, and G. A. Vargo, 374–76. Florida Fish and Wildlife Conservation Commission, Florida Institute of Oceanography, and Intergovernmental Oceanographic Commission of UNESCO, 2004.

Wait, P. "Looming Crisis: Falling Waterfowl Hunter Numbers Threaten the Future of Hunting and Conservation." *Delta Waterfowl.* Spring 2017. https://deltawaterfowl.org/wp-content/uploads/2017/03/LoomingCrisis.pdf. Accessed May 23, 2019.

Waycott, M., C. M. Duarte, T. J. B. Carruthers, R. J. Orth, W. C. Dennison, S. Olyarnik, A. Calladine, J. W. Fourqurean, K. L. Heck Jr., A. R. Hughes, G. A. Kendrick, W. J. Kenworthy, F. T. Short, and S. L. Williams. "Accelerating Loss of Seagrasses across the Globe Threatens Coastal Ecosystems." *Proceedings of the National Academy of Science* 106(30) (2009): 12377–81.

Weller, M. W. "Courtship of the Redhead (*Aythya americana*)." *The Auk* 84(4) (1967): 544–59.

———. "Distribution and Migration of the Redhead." *The Journal of Wildlife Management* 28(1) (1964): 64–103.

———. "Parasitic Egg Laying in the Redhead (*Aythya americana*) and Other North American Anatidae." *Ecological Monographs* 29(4) (1959): 333–65.

White, D. H., and D. James. "Differential Use of Fresh Water Environments by Wintering Waterfowl of Coastal Texas." *The Wilson Bulletin* 90(1) (1978): 99–111.

Wilkins, K., and E. G. Cooch. *Waterfowl Population Status, 1999*. U.S. Fish and Wildlife Publications Paper 404, 1999. Accessed at http://digitalcommons.unl.edu/usfwspubs/404.

Williams, B. K., M. D. Koneff, and D. A. Smith. "Evaluation of Waterfowl Conservation under the North American Waterfowl Management Plan." *The Journal of Wildlife Management* 63(2) (1999): 417–40.

Williams, C. S. "Migration of the Redhead from the Utah Breeding Grounds." *The Auk* 61(2) (1944): 251–59.

Withers, K. "Seagrass Meadows." In *The Laguna Madre of Texas and Tamaulipas*. Edited by J. W. Tunnell Jr. and F. W. Judd, 85–101. College Station: Texas A&M University Press, 2002: xxii, 346.

Woodin, M. C., T. C. Michot, and M. C. Lee. "Salt Gland Development in Migratory Redheads (*Aythya americana*) in Saline Environments on the Winter Range, Gulf of Mexico, USA." *Acta Zoologica Academiae Scientarum Hungaricae* 54 (Suppl. 1) (2008): 251–64.

Woodin, M. C., and G. A. Swanson. "Foods and Dietary Strategies of Prairie-nesting Ruddy Ducks and Redheads." *The Condor* 91 (1989): 280–87.

Zimpfer, N. L., W. E. Rhodes, E. D. Silverman, G. S. Zimmerman, and K. D. Richkus. *Trends in Duck Breeding Populations, 1955–2013*. Laurel, MD: U.S. Fish and Wildlife Service, Division of Migratory Bird Management, 2013: 25.

CHAPTER 5

Able, K. P. "Fall Migration in Coastal Louisiana and the Evolution of Migration Patterns in the Gulf Region." *The Wilson Bulletin* 84(3) (1972): 231–42.

Albanese, G., and C. A. Davis. "Characteristics Within and Around Stopover Wetlands Used by Migratory Shorebirds: Is the Neighborhood Important?" *The Condor: Ornithological Applications* 117 (2015): 328–40.

Andres, B. A., P. A. Smith, R. I. G. Morrison, C. L. Gratto-Trevor, S. C. Brown, and C. A. Friis. "Population Estimates of North American Shorebirds, 2012." *Wader Study Group Bulletin* 119(3) (2012): 178–94.

Arvin, J. "A Survey of Upper Texas Coast Critical Habitats for Migratory and Wintering Piping Plover and Associated Resident 'Sand Plovers.'" Texas Parks and Wildlife, Endangered Species Program, Grant No. TX E-95-R, 2008: 6.

Baker, A. J., P. M. Gonzalez, T. Piersma, L. J. Niles, I. de Lima Serrano do Nascimento, P. W. Atkinson, N. A. Clark, C. D. T. Minton, M. K. Peck, and G. Aarts. "Rapid Population Decline in Red Knots: Fitness Consequences of Decreased Refuelling Rates and Late Arrival in Delaware Bay." *Proceedings of the Royal Society of London*, B 271, 2004: 875–82. The Royal Society doi 10.1098/rspb.2003.2663

Bart, J., S. Brown, B. Harrington, and R. I. Guy Morrison. "Survey Trends of North American Shorebirds: Population Declines or Shifting Distributions?" *Journal of Avian Biology* 38(1) (2007): 73–82.

Berg, R. "Tropical Cyclone Report Hurricane Ike." National Hurricane Center. http://www.nhc.noaa.gov/data/tcr/AL092008_Ike.pdf. 14, 2009. Accessed December 24, 2015.

Bettencourt, S. "A History of the High Island Sanctuaries." Unpublished report by Houston Audubon Society, 1996: 10.

Bildstein, K. L., J. Zalles, J. Ottinger, and K. McCarty. "Conservation Biology of the World's Migratory Raptors: Status and Strategies." In *Raptors at Risk*. Edited by R. Chancellor and D. B.-U. Meyburg, 573–90. Hancock House, World Working Group on Birds of Prey and Owls, 2000.

Blake, E. S. "Tropical Cyclone Report Hurricane Humberto." http://www.nhc.noaa.gov/data/tcr/AL092007_Humberto.pdf. 16, 2007. Accessed Dec. 24, 2015.

Blake, E. S., C. W. Landsea, and E. J. Gibney. "The Deadliest, Costliest, and Most Intense United States Tropical Cyclones from 1851 to 2010 (and Other Frequently Requested Hurricane Facts)." NOAA Technical Memorandum NWS NHC-6. Miami: National Weather Service, National Hurricane Center, 2011: 47.

Boland, J. M. "An Overview of the Seasonal Distribution of the North American Shorebirds." *Wader Study Group Bulletin* 62 (1991): 39–43.

Brigance, J. "Bolivar Flats Important Resting Stop for Millions of

Birds." *Galveston Daily News.* March 22, 1993.

Britton, J. C., and B. Morton. *Shore Ecology of the Gulf of Mexico.* Austin: University of Texas Press, 1989: viii, 387.

Brown, S., C. Hickey, B. Harrington, and R. Gill, eds. *The U.S. Shorebird Conservation Plan.* 2nd ed. Manomet, MA: Manomet Center for Conservation Sciences, 2001: 64.

Brush, T. *Nesting Birds of a Tropical Frontier: The Lower Rio Grande Valley of Texas.* College Station: Texas A&M University Press, 2005: xiv, 245.

Buss, I. O. "Bird Detection by Radar." *The Auk* 63 (1946): 315–18.

Cestari, C. "Foraging Behavior of Hudsonian Godwit (*Limosa haemastica*) (Charadriiformes, Scolopacidae) in Human-disturbed and Undisturbed Occasions in the Atlantic Coast of Brazil." *Revista Brasileira de Ornitologia* 19(4) (2011): 535–38.

Chesser, R. T., K. J. Burns, C. Cicero, J. L. Dunn, A. W. Kratter, I. J. Lovette, P. C. Rasmussen, J. V. Remsen, Jr., D. F. Stotz, and K. Winker. 2019. Check-list of North American Birds (online). American Ornithological Society. http://checklist.aou.org/taxa

Cohen, E. B., Z. Németh, T. J. Zenzal Jr., K. L. Paxton, R. Diehl, E. H. Paxton, and F. R. Moore. "Spring Resource Phenology and Timing of Songbird Migration across the Gulf of Mexico." In *Phenological Synchrony and Bird Migration: Changing Climate and Seasonal Resources in North America.* Edited by E. M. Wood and J. L. Kellermann, 63–82. Studies in Avian Biology (no. 47), Boca Raton, FL: CRC Press, 2015.

Contreras, S. "Temporal and Spatial Patterns of Bird Migration in the Lower Gulf Region." PhD diss., Texas A&M University–Kingsville, 2010: 21.

Davis, C. A., and L. M. Smith. "Ecology and Management of Migrant Shorebirds in the Playa Lakes Region of Texas." *Wildlife Monographs* 140 (1998): 3–45.

Donaldson, G., C. Hyslop, G. Morrison, L. Dickson, and I. Davidson, eds. *Canadian Shorebird Conservation Plan.* Environment Canada, Canadian Wildlife Service, 2000: ii, 34.

eBird. http://ebird.org/content/ebird/, 2016. Accessed May 9, 2015.

Eldridge, Jan, "13.2.14. Management of Habitat for Breeding and Migrating Shorebirds in the Midwest." *Waterfowl Management Hand-*

book. Paper 11, 1992. http://digitalcommons.unl.edu/icwdmwfm/11.

Eubanks, T. L., Jr., R. A. Behrstock, and R. J. Weeks. *Birdlife of Houston, Galveston, and the Upper Texas Coast.* College Station: Texas A&M University Press, 2006: xii, 287.

Faaborg, J., R. T. Holmes, A. D. Anders, K. L. Bildstein, K. M. Dugger, S.A. Gauthreaux Jr., P. Heglund, K. A. Hobson, A. E. Jahn, D. H. Johnson, S. C. Latta, D. J. Levy, P. P. Marra, C. L. Merkord, E. Nol, S. L. Rothstein, T. W. Sherry, T. S. Sillett, F. R. Thompson III, and N. Warnock. "Recent Advances in Understanding Migration Systems of New World Land Birds." *Ecological Monographs* 80(1) (2010): 3–48.

Foster, C. R., A. F. Amos, and L. A. Fuiman. "Trends in Abundance of Coastal Birds and Human Activity on a Texas Barrier Island over Three Decades." *Estuaries and Coasts* 32 (2009): 1079–89.

Fotheringham, N. *Beachcomber's Guide to Gulf Coast Marine Life.* Houston: Gulf Publishing Company, 1980: xi, 124.

Galveston County, TX. *Blueprint for Bolivar.* Office of Emergency Management. www.gcoem.com, 2009: 103. Accessed July 4, 2016.

The Galveston Daily News. "Bolivar Flats Bird Sanctuary Formally Opens." April 26, 1992.

The Galveston Daily News. "The Bolivar Point Plan." January 16, 1893.

The Galveston Daily News. Editorial: "Bolivar Flats." April 5, 1993.

The Galveston Daily News. "Petition Is Started against Closing of Beach." February 3, 1985.

Gauthreaux, S. A., Jr. "Behavioral Responses of Migrating Birds to Daylight and Darkness: A Radar and Direct Visual Study." *The Wilson Bulletin* 84(2) (1972): 136–48.

———. "Historical Perspectives. Bird Migration: Methodologies and Major Research Trajectories (1945–1995)." *The Condor* 98 (1996): 442–53.

———. "Neotropical Migrants and the Gulf of Mexico: The View from Aloft." In *A Gathering of Angels: Migrating Birds and Their Ecology.* Edited by K. P. Able, 27–49. Ithaca and London: Comstock Books, 1999: 193.

Gauthreaux, S. A., Jr., and C. G. Belser. "Radar Ornithology and the Conservation of Migratory Birds." USDA Forest Service Gen. Tech. Rep. PSW-GTR-191, 2005: 871–75.

Gauthreaux, S. A., Jr., C. G. Belser, and C. M. Welch. "Atmospheric Trajectories and Spring Bird Migration across the Gulf of Mexico," n.d. https://www.researchgate.net/profile/Sidney_Gauthreaux/publication/226571918_Atmospheric_trajectories_and_spring_bird_migration_across_the_Gulf_of_Mexico/links/0912f51016019e8e39000000.pdf. Accessed May 10, 2016.

Gill, F. B. *Ornithology*. New York: W.H. Freeman, 1990: x, 660.

Haig, S. M., C. L. Ferland, F. J. Cuthbert, J. Dingledine, J. P. Goossen, A. Hecht, and N. McPhillips. "A Complete Species Census and Evidence for Regional Declines in Piping Plovers." USGS Staff— Published Research. Paper 683. http://digitalcommons.unl.edu/usgsstaffpub/683, 2005: 16. Accessed February 27, 2016.

Harrington B, A., C. Picone, L. Resende, and F. Leeuwenberg. "Hudsonian Godwit (*Limosa haemastica*) Migration in Southern Argentina." *Wader Study Group Bulletin* 67 (1993): 41–44. https://sora.unm.edu/sites/default/files/journals/iwsgb/n067/p00041-p00043.pdf. Accessed July 4, 2016.

Houston Audubon Society. "Bird Checklist, Bolivar Flats Shorebird Sanctuary, Horseshoe Marsh Bird Sanctuary, Mundy Bay Marsh Sanctuary," n.d.

Houston Audubon Society. *Birds of Bolivar*. Sausalito, CA: Mitch Waite Group, 2006: 305.

Houston Audubon Society. "History of the Bolivar Flats Shorebird Sanctuary." Unpublished report, 2015: 3.

Houston Audubon Society/Texas General Land Office. "Bolivar Flats Protection Plan," 2010: 14.

Houston Outdoor Nature Club, Ornithology Group. *A Birder's Checklist of the Upper Texas Coast*, 9th ed., May 2008.

Howe, M. A., P. H. Geissler, and B. A. Harrington. "Population Trends of North American Shorebirds Based on the International Shorebird Survey." *Biological Conservation* 49 (1989): 185–99.

King, K. A. "Bird Mortality, Galveston Island, Texas." *The Southwestern Naturalist* 21(3) (1976): 414.

Kirkpatrick, J. "Bird Sanctuary Plan Delayed to Let Opposing Sides Settle Issue." *Galveston Daily News*. February 12, 1985.

——. "County Approves Pilings, Cable at Bird Sanctuary." *Galveston Daily News*. December 30, 1986.

Kraus, N. C., and L. Lin. "Hurricane Ike along the Upper Texas Coast: An Introduction." *Shore & Beach* 77(2) (2009): 3–8.

Langin, K. M., P. P. Marra, Z. Nemeth, T. K. Kyser, and L. M. Ratcliff. "Breeding Latitude and Timing of Spring Migration in Songbirds in Crossing the Gulf of Mexico." *Journal of Avian Biology* 40 (2009): 309–16.

Lowery, J. H., Jr. "Evidence of Trans-gulf migration." *The Auk* 63 (1946): 175–211.

——. "Trans-gulf Migration of Birds and the Coastal Hiatus." *The Wilson Bulletin* 57(2) (1945): 92–121.

Massey, R. "Bird Society Buys 615 Acres on Bolivar." *Galveston Daily News.* December 6, 2001.

Myers, J. P. "Conservation of Migrating Shorebirds: Staging Areas, Geographic Bottlenecks, and Regional Movements." *American Birds* 37(1) (1983): 23–25.

NASA. "Post-Ike Aerial Photograph of High Island." http://earthobservatory.nasa.gov/IOTD/view.php?id=9107. Accessed May 16, 2016,

Nash, J. P. "Texas Granites." University of Texas Bulletin No. 1725, 1917: 13.

National Geographic Society. *Field Guide to the Birds of North America.* 3rd ed. Washington, DC, 1999: 480.

National Hurricane Center. "Costliest U.S. Tropical Cyclones Tables Updated." https://www.nhc.noaa.gov/news/UpdatedCostliest.pdf, 2018. Accessed May 24, 2019.

National Oceanic and Atmospheric Administration, Office of Coastal Management. "Hurricane Costs." https://coast.noaa.gov/states/fast-facts/hurricane-costs.html, 2019. Accessed May 23, 2019.

Neill, R. L. "Recent Trends in Shorebird Migration for North-central Texas." *The Southwestern Naturalist* 37(1) (1992): 87–88.

The New York Times. "A deep gulf harbor." February 9, 1890. http://query.nytimes.com/mem/archive-free/pdf?_r=1&res=9501E0D-7143BE533A2575AC0A9649C94619ED7CF&oref=slogin. Accessed July 4, 2016.

Nicholls, J. L., and G. A. Baldassarre. "Winter Distribution of Piping Plovers along the Atlantic and Gulf Coasts of the United States." *The Wilson Bulletin* 102(3) (1990): 400–412.

Niven, D. K., and G. S. Butcher. "Status and Trends of Wintering

Coastal Species along the Northern Gulf of Mexico." *American Birds: The 111ᵗʰ Christmas Count.* http://mag.audubondrupal.org/sites/default/files/documents/ab_111_trends_gulf.pdf, 2011, 12–19. Accessed June 30, 2016.

North American Bird Conservation Initiative. "Partners in Flight." *State of North America's Birds 2016.* http://www.stateofthebirds.org/2016/wp-content/uploads/2016/05/SoNAB-ENGLISH-web.pdf, 2016. Accessed July 9, 2016.

Oberholser, H. C., and E. B. Kinkaid Jr., eds. *The Bird Life of Texas,* vol. 1. Austin: University of Texas Press, 1974: xxviii, 530.

Ortego, B., and M. Ealy. "2009 Winter Texas Gulf Coast Aerial Shorebird Survey." *Bulletin of the Texas Ornithological Society* 43(1-2) (2010): 1–10.

Partners in Flight Science Committee. "Population Estimates Database, Version 2013." http://rmbo.org/pifpopestimates. Accessed March 23, 2016,

Rappole, J. H. *The Avian Migrant: The Biology of Bird Migration.* New York: Columbia University Press, 2013: 435.

Rappole, J. H., B. Helm, and M. A. Ramos. "An Integrative Framework for Understanding the Origin and Evolution of Avian Migration." *Journal of Avian Biology* 34(1) (2003): 124–28.

Sauer, J. R., J. E. Hines, J. E. Fallon, K. L. Pardieck, D. J. Ziolkowski Jr., and W. A. Link. *The North American Breeding Bird Survey, Results and Analysis 1966–2013. Version 01.30.2015.* Laurel, MD: USGS Patuxent Wildlife Research Center, 2014.

Senner, N. R. *Conservation Plan for the Hudsonian Godwit.* Version 1.1. Manomet, MA: Manomet Center for Conservation Science, 2010: 63.

———. "The Status and Conservation of Hudsonian Godwits (*Limosa haemastica*) during the Non-breeding Season." *Ornitologia Neotropical* 19 Suppl. (2008): 623–31.

Shackelford, C. E., E. R. Rozenburg, W. C. Hunter and M. W. Lockwood. *Migration and the Migratory Birds of Texas: Who They Are and Where They Are Going.* Texas Parks and Wildlife PWD BK W7000-511 (11/05), 2005: 34.

Simons, T. R., F. R. Moore, and S. A. Gauthreaux. "Mist Netting Trans-Gulf Migrants at Coastal Stopover Sites: The Influence of Spatial and Temporal Variability on Capture Data." *Studies in Avian Biology* 29 (2004): 135–43.

Skagen, S. K., and F. L. Knopf. "Migrating Shorebirds and Habitat Dynamics at a Prairie Wetland Complex." *The Wilson Bulletin.* 106(1) (1994): 91–105.

Skagen, S. K., P. B. Sharpe, R. G. Waltermire, and M. B. Dillon. *Biogeographical Profiles of Shorebird Migration in Mid-continental North America.* Biological Science Report USGS/BRD/BSR — 2000-0003. Denver, CO: U.S. Government Printing Office, 1999: 167.

Stevenson, J. O. "Bird Notes from the Texas Coast." *The Wilson Bulletin* 65(1) (1953): 42–43.

Stevenson, H. M. "The Relative Magnitude of the Trans-gulf and Circum-gulf Spring Migrations." *The Wilson Bulletin* 69(1) (1957): 39–77.

Terborgh, J. *Where Have All the Birds Gone?* Princeton, NJ: Princeton University Press, 1989: xvi, 207.

Thomas-Hargis, K. "Bird Sanctuary Sought." *Galveston Daily News.* November 28, 1984.

U.S. Census Bureau. "Census of Population and Housing, Population and Housing Unit Counts," 2010. CPH-2-45, Texas. Washington, DC: Government Printing Office, 2012.

U.S. Fish and Wildlife Service. "Biological Opinion on Impacts of Project on Piping Plover Habitat." Consultation No. 21430-2011-F-0281. December 22, 2011, 60.

U.S. Fish and Wildlife Service. "Birds of Anahuac National Wildlife Refuge," 2003. https://www.fws.gov/uploadedFiles/Anahuac_Bird_List_508.pdf.

U.S. Fish and Wildlife Service. "High Priority Shorebirds — 2004." Arlington, VA: U.S. Shorebird Conservation Plan MBSP 4107, 2004: 5.

U.S. Geological Survey. "North American Breeding Bird Survey." https://www.pwrc.usgs.gov/bbs/, 2016. Accessed September 12, 2016.

Western Hemisphere Shorebird Reserve Network. http://www.whsrn.org/western-hemisphere-shorebird-reserve-network, 2016. Accessed September 12, 2016.

Williams, A. M., R. A. Feagin, W. K. Smith, and N. L. Jackson. "Ecosystem Impacts of Hurricane Ike on Galveston Island and Bolivar

Peninsula: Perspectives of the Coastal Barrier Island Network (CBIN)," 2009: 17. http://users.wfu.edu/smithwk/CBIN/Manuscript1.pdf 2009. Accessed February 17, 2016.

Williams, G. G. "Do Birds Cross the Gulf of Mexico in Spring?" *The Auk* 62 (1945): 98–110.

Williams, G. G. "The Nature and Causes of the 'Coastal Hiatus.'" *The Wilson Bulletin* 62(4) (1950): 175–82.

Withers, K. "Shorebird Use of Coastal Wetland and Barrier Island Habitat in the Gulf of Mexico." *The Scientific World Journal* 2 (2002): 514–36.

Zane, D. F., T. M. Bayleyegn, J. Hellsten, R. Beal, C. Beasley, T. Haywood, D. Wiltz-Beckham, A. F. Wolkin. "Tracking Deaths Related to Hurricane Ike, Texas, 2008." *Disaster Med Public Health Prep* 5(1) (2011): 23–28. https://www.ncbi.nlm.nih.gov/pubmed/21402823. Accessed June 16, 2019.

Zink, R. M. "Towards a Framework for Understanding the Evolution of Avian Migration." *Journal of Avian Biology* 33(4) (2002): 433–36.

Index

the Houston
Museum *of*
natural science

Publication of this book was made
possible by the generous support of the
Houston Museum of Natural Science.